Lecture Notes in Computer Science

Lecture Notes in Computer Science

Edited by G. Goos and J. Hartmanis

413

Reiner Lenz

Group Theoretical Methods in Image Processing

Springer-Verlag

Berlin Heidelberg New York London Paris Tokyo Hong Kong

Author

Reiner Lenz
Department of Electrical Engineering
Linköping University
S-58183 Linköping, Sweden

CR Subject Classification (1987): I.4.5–7, I.5.1, I.5.4

ISBN 978-3-540-52290-4   ISBN 978-3-540-46947-6 (eBook)
DOI 10.1007/978-3-540-46947-6

Printing and binding: Druckhaus Beltz, Hemsbach/Bergstr.
2145/3140-543210 – Printed on acid-free paper

Foreword

These Lecture Notes are based on a series of lectures I gave at the Linköping University Department of Electrical Engineering in 1988. In these lectures I tried to give an overview of the theory of representation of compact groups and some applications in the fields of image science and pattern recognition.

The participants in the course had a masters degree in electrical engineering and no deeper knowledge in group theory or functional analysis. The first chapter is thus used to introduce some basic concepts from (algebraic) group theory, topology, functional analysis and topological groups. This chapter contains also some of the main results from these fields that are used in the following chapters. This chapter cannot replace full courses in algebra, topology or functional analysis but it should give the reader an impression of the main concepts, tools and results of these branches of mathematics. Some of the groups that will be studied more extensively in the following chapters are also introduced in this chapter.

The central idea of group representations is then introduced in Chapter 3. The basic theme in this chapter is the study of function spaces that are invariant under a group of transformations such as the group of rotations. It will be demonstrated how the algebraic and topological properties of the group determine the structure of these invariant spaces: if the transformation group is commutative, it will be shown that the minimal, invariant spaces are all one-dimensional and if the group is compact, then these spaces are all finite-dimensional. This result explains, for example, the unique role of the complex exponential function. In this chapter we concentrate on the derivation of qualitative information about the invariant function spaces but (with the exception of the complex exponential function that is connected to the group of real numbers **R** and the group of 2-D rotations) no such function spaces are actually constructed. The construction of these invariant function spaces for some important groups is demonstrated in Chapter 4. There we use methods from the theory of Lie groups to find the functions connected to certain important transformation groups. These groups include the scaling group, the 2-D and 3-D rotation groups, the group of rigid motions and the special linear group of the 2×2 matrices with complex entries and determinant equal to one.

In Chapter 5 we conclude the mathematical part of these lectures by demonstrating how the concept of a Fourier series can be generalized. The basic idea behind this generalization is the following: Consider a periodic function $f(\phi)$ with period 2π. Usually we think of such a function as a (complex-valued) function on an interval or a circle. In the new interpretation we view ϕ as the angle of a rotation and f becomes thus a function defined on the group of 2-D rotations. The decomposition of f into a Fourier series means thus that a function on the group of 2-D rotations can be decomposed into a series of complex exponentials. In Chapter 5 we demonstrate that a similar decomposition is possible for functions defined on compact groups.

In Chapter 6 it is then demonstrated how the results from the previous chapters can be used to investigate and solve problems in image science. Our approach to solve these problems is based on the observation that many problems in image science are highly regular or symmetric. We investigate how these symmetries can be used to design problem solving strategies that can make use of this information.

First it is demonstrated how one can find eigenvectors of operators that commute with the group transformation. As important examples it is shown that eigenvalue problems connected to the group of 2-D rotations lead to the complex exponentials as solutions, whereas operators that commute with 3-D rotations have eigenvectors connected to the spherical harmonics. Examples of such operators are the Laplacian and the Fourier Transform.

As an example we then give a short overview of series expansions methods in computer aided tomography. Here we use the fact that the scanning process is rotationally symmetric. The functions used in the reconstruction process are thus tightly coupled to the rotation groups.

Another application deals with the generalization of the well-known edge-detection problem. We generalize it to the case where the interesting pattern is an arbitrary function and where the images are of dimension two or higher. We only assume that the transformation rule can be defined in a group theoretical sense. Under these conditions we describe a method that allows us to solve the pattern detection problem. Moreover we will show that the method derived is optimal in a certain sense. It will also be demonstrated why the derived filter functions are often found as solutions to optimality problems and why they have a number of other nice properties. In the rotational case for example it will be shown that the filter functions are eigenfunctions of the Laplacian and the Fourier Transform. Another result shows that group representations naturally turn up in the Karhunen Loeve expansion which makes these results interesting for applications in image coding.

We will see that most of the problems mentioned can be formulated in terms of so called compact groups. These are groups that do not have too many elements. We will also, briefly, study the easiest example of a non-compact, non-commutative group. This is the group of all complex 2×2 matrices with determinant one. It will be demonstrated how this group can be used in the analysis of camera motion.

In the last example we study some problems from the theory of neural networks. Here one investigates so-called basic units. These basic units are a model of one or a small number of nerve-cells. One approach to investigate and design such basic units is to view such a unit as a device that correlates the incoming signal with its internal state. The correlation result is then send to other units in the network and the internal state is updated according to some rule. In the last section we assume that the unit updates its internal state in such a way that the computed result is maximal in a mean-squared sense. We then investigate what happens when the basic unit is trained with a set of input functions that are transformed versions of a fixed pattern. As example we consider the cases were the input signals are rotated edge or line patterns. We show that in these cases the stable states of the unit are connected to the representations of the group of transformations that generated the input patterns from the fixed pattern.

Acknowledgements

I want to thank Prof. P. E. Danielsson who introduced me to the invariance problems in image processing. I also have to thank him for his support and many stimulating discussions.

A. Grönqvist did most of the experiments with the MR-images and the adaptive filter systems were simulated by M. Österberg.

The MR volume was supplied by Dr. Laub, Siemens AG, Erlangen, W-Germany.

The financial support from the Swedish Board for Technical Development is gratefully acknowledged.

Linköping, November 1989

Reiner Lenz

Contents

Chapter 1

Introduction

In these notes we will try to give an introduction to the theory of group representations and we will demonstrate the usefulness of this theory by investigating several problems from image science. This group theoretical approach was motivated by the observation that many problems in image processing are very regular. In this introduction we will, as an example, study the well-known edge-detection problem for 2-D images.

1.1 Edge Detection

In a first attempt we can formulate this problem as follows:

A point in an image is an *edge point* if there is a line through this point such that all points on one side of this line are dark and all points on the other side are light. The problem to find all edge points in a given image is called the *edge detection problem*.

This definition of an edge point contains one degree of freedom since we did not specify the direction of the line that separates the black and the white points. There is thus in infinite number of different types of edge points: one type of edge point for each direction of the edge. We reformulate the edge detection problem as follows:

Define L_ψ as the line that goes through the origin of the coordinate system and has an angle ψ with the x-axis. By p_ψ we denote the function that has the value one on the left side of L_ψ and the value zero on the other side. The set of step edges, denoted by $\mathcal{E}$, is the set of all functions p_ψ with $0 \leq \psi < 2\pi$.

Given a point P in the image with gray value distribution f in a neighborhood we want to decide if $f \in \mathcal{E}$ or $f \notin \mathcal{E}$. In the first case we say that P is an edge point.

This formulation of the problem reveals that the set of edge patterns $\mathcal{E}$ is not just an arbitrary collection of patterns. On the contrary, it is highly structured, or symmetric, since we can generate all elements in this set by rotating a fixed pattern around the origin. The fact that the set of patterns that we want to detect is very regular is so important that we introduce a special concept to describe this type of regularity.

1.2 Invariant Pattern Classes

In the general case we consider a set of functions $\mathcal{C}$. The elements in $\mathcal{C}$ are the patterns we want to detect in the image. The fundamental pattern recognition problem is thus to decide if a given pattern p is an element of $\mathcal{C}$ or not.

Without any restriction on $\mathcal{C}$ we can solve this problem only by comparing p with all elements in the pattern class $\mathcal{C}$. However, the situation changes drastically if $\mathcal{C}$ is structured. The interesting structures of $\mathcal{C}$ are based on two properties:

Domain: First there is a common domain for all functions in $\mathcal{C}$, i.e. a common set on which all the pattern in $\mathcal{C}$ are defined. In the edge-detection problem this domain is the unit disk.

Transformation: Furthermore the patterns in $\mathcal{C}$ are all coupled by a transformation rule. This means that we can generate all patterns in the class by applying the transformation rule to a fixed pattern in the class. In the case of the step edges the transformation rule is defined in terms of rotations.

The essential part in this construction is the transformation rule. This rule links all the patterns in the pattern class together. We will sometimes also say that all the patterns in the class are essentially the same.

One of the main goals of these lectures is the exact definition and the investigation of this type of pattern class.

1.3   Pattern Detection

Now we return to the edge detection problem and we demonstrate that there is a connection between the description of the pattern class and the detection of these patterns in an image.

We saw that the step edges were functions of the form $p_\psi(r,\varphi)$, where (r,φ) are the polar coordinates of a point. We ignore the different character of the variables r, φ and ψ and write $p(r,\varphi,\psi)$. If we fix the spatial variables r and φ then we can consider p as a function of the transformation variable ψ alone.

A Fourier decomposition of p leads to the following *weighted group averages*

$$P_k(r,\varphi) = \int_0^{2\pi} p(r,\varphi,\psi)e^{ik\psi}\,d\psi \qquad (1.1)$$

Note that this defines a new set of (complex-valued) patterns defined on the same domain. These patterns define the pattern class completely since we can recover the original patterns by forming linear combinations of the new patterns P_k:

$$p(r,\varphi,\psi) = \sum a_k e^{ik\psi} P_k(r,\varphi) \qquad (1.2)$$

This decomposition of the patterns has the advantage that the spatial variables r and φ and the transformation variable ψ are separated. Furthermore we see that the different components $e^{ik\psi}P_k(r,\varphi)$ in the decomposition transform nicely when we go from $p(r,\varphi,\psi_0)$ to $p(r,\varphi,\psi_0+\psi_1)$: every component is simply multiplied by a complex factor.

Now assume that we have an unknown pattern p_u and we want to decide if p_u is a member of the pattern class. To solve this problem we compute again P_k and find:

$$P_k(r,\varphi) = \int_0^{2\pi} p(r,\varphi,\psi)e^{ik\psi}\,d\psi \qquad (1.3)$$

$$= \int_0^{2\pi} p(r,\varphi+\psi,0)e^{ik\psi}\,d\psi \qquad (1.4)$$

$$= e^{-ik\varphi} \int_0^{2\pi} p(r, \psi, 0) e^{ik\psi} \, d\psi \qquad (1.5)$$

The weighted average over the pattern class is thus (up to a constant complex factor of magnitude one) equal to the *weighted spatial average* over one pattern in the class.

We can use this relation to decide if a pattern p_u is equal to one pattern $p(r, \varphi, \psi)$ in the pattern class by doing the following computations:

- First we compute the weighted spatial averages from the prototype pattern. These averages are, up to a complex factor, equal to the weighted class averages over the pattern class. The result are complex functions $P_k(r, 0)$.

- Then we compute the spatial weighted averages over the unknown pattern. As result we get complex functions $U_k(r)$.

- If now $P_k(r, 0) = e^{ik\alpha} U_k(r)$ for a constant α and all k then we know that the unknown pattern was a pattern in the given pattern class.

We found thus that the weighted class averages and the weighted spatial averages were very useful in the description and the analysis of the pattern class. The usefulness of this approach depended of course on the clever selection of the weight function $e^{ik\psi}$ and one of the main purposes of these notes is the construction of similar useful weight functions for a number of different types of invariant pattern classes.

Chapter 2

Preliminaries

In this chapter we collect some of the basic facts from algebra, topology, functional analysis and the theory of topological groups. This chapter cannot replace full courses in these topics and it is not expected that a reader unfamiliar with these fields can master them on first reading. We include this short summary to make these notes more or less self contained and to give the reader a feeling for the problems investigated in these branches of mathematics. In this chapter we will also introduce some basic notations and some important results for easier reference. On first reading the reader may browse through this chapter and come back to it later when reading the following chapters. For a more detailed treatment of these topics it is necessary to consult the standard literature.

2.1 Topology and Functional Analysis

We introduce first some basic notations and generalize then the concept of a neighborhood of a point by introducing topological spaces. Such a space is a set together with a system of subsets of the original set. The subsets in this system satisfy certain conditions that characterize open subsets in the usual meaning. We call the subsets in this system open subsets. We will not develop topology in any detail but we will only introduce the concepts of compact and locally compact sets and continuous mappings between topological spaces.

Then we give a short overview of Lebesgue integration that will be used later when we introduce invariant integration over groups. Finally we introduce some basic definitions from functional analysis, especially Hilbert spaces and operators between them.

This section collects only the most important definitions and some results from these areas and the interested reader may wish to consult the standard literature for further information.

2.1.1 Some Notations

By $\emptyset$ we denote the *empty set*, by $\{x, y, z\}$ the set with the elements x, y and z and by $\{x | c(x)\}$ the set of all elements which satisfy the condition $c(x)$.

The notation $f : A \to B; a \mapsto b = f(a)$ describes the *function* f that maps the set A into the set B. f maps the element $a \in A$ into the element $b = f(a)$. If $X \subset B$ then we denote by $f^{-1}(X)$ the set

$$f^{-1}(X) = \{a \in A | f(a) \in X\} \tag{2.1}$$

By $f(A)$ or $im(f)$ we denote the set of all elements in A that are mapped into X under f:

$$f(A) = \{b \in B | b = f(a) \text{ for some } a \in A\}. \tag{2.2}$$

We call $im(f)$ the *image* of A under f. If $f : A \to B$ and $g : B \to C$ then we denote by $g \circ f$ the *composition* of f and g defined as:

$$g \circ f : A \to C; a \mapsto g(f(a)). \tag{2.3}$$

A map is called *injective* if $f(a) = f(b)$ implies $a = b$. An injective function maps different elements in A to different elements in B. A map is called *surjective* if for all $b \in B$ there is an element $a \in A$ such that $f(a) = b$, or $im(f) = B$. A map is called *bijective* if it is injective and surjective.

If E_α is a collection of sets, where α runs over some index set A, then we define by $\bigcup_{\{\alpha \in A\}} E_\alpha$ the *union* of all the sets E_α and by $\bigcap_{\{\alpha \in A\}} E_\alpha$ their *intersection*. If A and B are two sets then we define their *product* $A \times B$ as the set $\{(a, b) | a \in A, b \in B\}$. The *Kronecker symbol* δ_{ij} is defined as the function which has the value 1 if $i = j$ and 0 if $i \neq j$.

Finally we recall the notation of an *equivalence relation*. Assume A is a set. Then we say that the subset X of $A \times A$ defines a equivalence relation if the following conditions are satisfied:

- $(a, a) \in X$ for all $a \in A$

- If $(a, b) \in X$ and $(b, c) \in X$ then $(a, c) \in X$

- If $(a, b) \in X$ then $(b, a) \in X$

If X defines an equivalence relation and $(a, b) \in X$ then we write $a \equiv b$. The set $E(a) = \{b \in A | a \equiv b\}$ is called an *equivalence class* and it is easy to show that an equivalence relation partitions A in the sense that $\bigcup_{a \in A} E(a) = A$ and $E(a) \bigcap E(b) = E(a)$ or $E(a) \bigcap E(b) = \emptyset$.

2.1.2 Topological Spaces

In the first definition we generalize the concept of open sets for arbitrary spaces.

Definition 2.1   1. A set X together with a family $\mathcal{U} = \{U\}$ of subsets $U \subset X$ is called a *topological space* if it satisfies the following conditions:

- $\emptyset \in \mathcal{U}$ and $X \in \mathcal{U}$.

- The union of any family of sets in $\mathcal{U}$ belongs to $\mathcal{U}$.

- The intersection of any finite number of sets in $\mathcal{U}$ belongs to $\mathcal{U}$.

2. The elements of $\mathcal{U}$ are called the *open sets* of the topological space X.

3. The family $\mathcal{U}$ is called the *topology* of X.

4. A set $\mathcal{V} = \{V\}$ of elements $V \in \mathcal{U}$ is called a *basis for the topology* $\mathcal{U}$ if every element $U \in \mathcal{U}$ is the union of certain elements $V \in \mathcal{V}$.

We note that a topological space consists of two elements, the set X and the topology $\mathcal{U}$. A topological space should therefore be denoted by the pair $(X,\mathcal{U})$, we will however mainly speak of the topological space X assuming that the topology is clear from the context. We have to denote the topology explicitly if we consider two different topologies on the same set X.

Examples 2.1   1. Take as $X = \mathbf{R}$ the set of real numbers. If we take the open intervals as the basis of the topology then we get the usual topology on $\mathbf{R}$.

2. Take any set X and define $\mathcal{U} = \{\emptyset, X\}$. This is the coarsest topology on X.

3. Take any set X and define $\mathcal{U}$ as the set of all subsets of X. This is the finest topology on X. It is called the *discrete topology*.

With the help of a topology it is possible to generalize the concept of a neighborhood of a point:

Definition 2.2 If $x \in X$ is a point in X, then we call any open set $U \in \mathcal{U}$ containing x a *neighborhood of the point* x. Neighborhoods of $x \in X$ will often be denoted by $U(x)$.

With this definition we find easily the familiar characterization of an open set:

Theorem 2.1 A subset $M \subset X$ is open if and only if every point $x \in M$ has a neighborhood $U(x)$ such that $U(x) \subset M$.

Note that the last statement is a theorem and not the definition of an open set. One construction which will be frequently used in the following is the definition of a topology on a subset $Y \subset X$. We define:

Definition 2.3 Assume $(X,\mathcal{U})$ is a topological space and $Y \subset X$. Then we define a topology $\mathcal{U}'$ on Y by using the intersections $Y \cap U$ as open sets on Y:

$$\mathcal{U}' = \{Y \cap U | U \in \mathcal{U}\}.$$

Closed sets are as usual defined as the complements of the open sets:

Definition 2.4   1. A subset $C \subset X$ is called a *closed subset* if the complement $X - C$ is open, i.e. $X - C \in \mathcal{U}$.

2. The intersection of all closed subsets containing $M \subset X$ is called the *closure of* M. It is denoted by $\overline{M}$.

3. An element in $\overline{M}$ is called a *limit point* of M.

The definition of a compact set is motivated by the well-known Heine-Borel theorem:

Definition 2.5   1. A family $\{O\}$ of subsets of X is called a *cover* if the union is equal to $X : X = \bigcup O$. If all sets O are open then we call it an *open cover*.

2. A topological space X is called *compact* if every open cover $\{O\}$ of X contains a finite cover $\{O_1, ..., O_n\}$.

3. A subset $Y \subset X$ is called *compact* if it is compact when regarded as a subspace of X.

4. A space X is called *locally compact* if every point $x \in X$ has a neighborhood whose closure is compact.

5. Assume $x_1, ..., x_n, ...$ is a sequence of elements in X. The point $x \in X$ is called a *limit of the sequence* if for every neighborhood $U(x)$ of x, there is a positive integer n_0 such that all elements $x_{n_0}, x_{n_0+1}, ...$ of the sequence belong to $U(x)$. We write $x = \lim_{n \to \infty} x_n$.

As a last type of topological spaces we introduce separated or Hausdorff spaces since we will always assume that the topological spaces are separated.

Definition 2.6 A topological space X is called *separated* or a *Hausdorff space* if every two points $x_1, x_2 \in X$ can be separated by disjoint neighborhoods. For every two points $x_1 \neq x_2$ we can thus find two neighborhoods $U(x_1), U(x_2)$ such that $U(x_1) \cap U(x_2) = \emptyset$.

Finally we will now define continuous mappings between arbitrary topological spaces:

Definition 2.7 Assume X and Y are two topological spaces and $f : X \to Y$ is a mapping.

1. Assume x_0 is a point in X and $y_0 = f(x_0)$. f is called *continuous at the point* $x_0 \in X$ if the inverse image $f^{-1}(V(y_0))$ of every neighborhood $V(y_0)$ contains some neighborhood $U(x_0)$ of $x_0 : U(x_0) \subset f^{-1}(V(y_0))$.

2. f is called *continuous* if f is continuous on every point in X.

3. f is called a *homeomorphism* if:

   - f is bijective and if
   - f and f^{-1} are continuous

For continuous mappings we find the following characterization:

Theorem 2.2   1. If $(X, \mathcal{U})$ and $(Y, \mathcal{V})$ are topological spaces and f is a function $f : X \to Y$, then f is continuous if and only if the inverse image of every open set in Y is open in X: $f^{-1}(V) \in \mathcal{U}$ for all $V \in \mathcal{V}$.

2. If f is a function $f : X \to Y$, then f is continuous if and only if the inverse image of every closed set in Y is closed in X.

3. Under a homeomorphism open sets are mapped onto open sets and closed sets are mapped onto closed sets.

2.1.3 Lebesgue Integrals

We summarize first some basic facts about Lebesgue integration since the existence of a special (group-invariant) integral is an essential result needed in the study of so-called locally-compact groups (see section 2.3.3). For a detailed treatment of measure and integration theory the reader may consult the literature (for example [17]).

Definition 2.8 1. Assume $\mathcal{S}^n$ is the set of all bounded intervals in $\mathbf{R}^n$. Then we say that a function ϕ is an *interval function* if $\phi : \mathcal{S}^n \to \mathbf{R}$.

2. An interval function ϕ is:

  - *monotone* if for all $I_1, I_2 \in \mathcal{S}^n$ with $I_1 \subset I_2 : \phi(I_1) \le \phi(I_2)$.

  - *additive* if for all $I_1, I_2, I \in \mathcal{S}^n$ with $I_1 \cap I_2 = \emptyset$ and
    $I_1 \cup I_2 = I : \phi(I) = \phi(I_1) + \phi(I_2)$

  - *regular* if for all $\epsilon > 0$ and all $I \in \mathcal{S}^n$ there is an open interval $I^* \supset I$ such that
    $\phi(I) \le \phi(I^*) < \phi(I) + \epsilon$.

3. A *measure* μ is a monotone, additive and regular interval function.

A typical interval function is the length (or the volume) of an interval: $\phi(I) = \phi([a, b]) = b - a$.

We use a measure on $\mathbf{R}^n$ to define the integral over those simple functions that are constant on intervals and have non-zero value only on a finite number of intervals:

Definition 2.9 Assume $I_1, ..., I_n$ is a finite number of intervals in $\mathbf{R}^n$ and μ is a measure. Assume further that $f : \mathbf{R}^n \to \mathbf{R}$ is a function that is constant on I_j where it has the value $c_j = f(I_j)$. Assume further that it is zero outside these intervals. Then we define the *Lebesgue integral* as $\int f \, d\mu = \int_{\mathbf{R}^n} f \, d\mu = \sum_{j=1}^n c_j \mu(I_j)$.

With the help of a limit operation this integral is then extended to a larger class of functions:

First we define $C_1(\mathbf{R}^n, \mu)$ as the set of functions for which there is a monotonically increasing sequence of interval functions f_n such that $f = \lim f_n$ and $\int f_n \, d\mu \le A$ for a constant A and all indices n. For the functions $f \in C_1(\mathbf{R}^n, \mu)$ we define $\int f \, d\mu = \lim \int f_n \, d\mu$. Then we define $C_2(\mathbf{R}^n, \mu) = \{f_1 - f_2 | f_i \in C_1(\mathbf{R}^n, \mu)\}$ and set $\int f \, d\mu = \int f_1 \, d\mu - \int f_2 \, d\mu$ for all $f \in C_2(\mathbf{R}^n, \mu)$.

Definition 2.10 A function f is called *integrable* if it is an element of $C_2(\mathbf{R}^n, \mu) : f \in C_2(\mathbf{R}^n, \mu)$.

To be correct one would have to change the previous definitions in such a way that two functions which have different values only on a set of measure zero are considered to be equal.

One of the main reasons for introducing a new integral (which is a generalization of the Riemann integral) is the ability to interchange integration and limit operations. One example of such a property and one of the main results of the theory is the following theorem:

Theorem 2.3 [Lebesgue] Assume f_n is a series of integrable functions and $f = \lim f_n$. Assume further that there is an integrable function g such that $|f_n| \leq g$ for all n. Then f is integrable and

$$\int \lim f_n \, d\mu = \int f \, d\mu = \lim \int f_n \, d\mu.$$

Finally we introduce the concept of a measurable function:

Definition 2.11 A function $f : \mathbf{R}^n \to \mathbf{R}$ is called *measurable* if there is a sequence of interval functions f_n such that the $\lim f_n = f$.

Note that measurable functions are not necessarily integrable since we required in the definition of the set C_1 that the integrals over the interval functions were bounded by a fixed constant. The definitions of measurable and integrable functions can easily be generalized to the case where the functions are complex valued. In this case one treats the function as a sum of two real functions $f = f_r + i \cdot f_i$ and defines $\int f \, d\mu = \int f_r \, d\mu + i \int f_i \, d\mu$.

With the help of the previous definitions we can introduce the spaces L^p defined as follows:

Definition 2.12 If $p > 0$ then:

1. $L^p(\mathbf{R}^n, \mu) = \{f : \mathbf{R}^n \to \mathbf{R} | f \text{ is measurable and} |f|^p \text{ is integrable}\}$
2. $L_{\mathbf{C}}^p(\mathbf{R}^n, \mu) = \{f : \mathbf{R}^n \to \mathbf{C} | f \text{ is measurable and} |f|^p \text{ is integrable}\}$

2.1.4   Functional Analysis

In this section we summarize some important concepts from functional analysis. This is only offered as a collection of results and not as an introduction to functional analysis. For such an introduction the reader is referred to the literature (see for example [41], [19], [53] and [10]).

We first introduce a norm as a map that generalizes the concept of length from geometry. Then we introduce a scalar product which is used to define the length of elements in a vector space and the angle between two elements in a vector space.

Definition 2.13   1. A *real vector space* is a set X together with two mappings $+ :$ $X \times X \to X$ and $\cdot : \mathbf{R} \times X \to X$ such that the following conditions are satisfied for all $x, y, z \in X$ and all $k, k_1, k_2 \in \mathbf{R}$:

(a) $(x + y) + z = x + (y + z)$

(b) There is a zero element $0 : x + 0 = x$ for all $x \in X$.

(c) For all $x \in X$ there is an inverse element $-x \in X : x - x = 0$.

(d) $x + y = y + x$

(e) $(k_1 + k_2) \cdot x = k_1 \cdot x + k_2 \cdot x$

(f) $k \cdot (x + y) = k \cdot x + k \cdot y$

(g) $(k_1 k_2) \cdot x = k_1 \cdot (k_2 \cdot x)$

(h) $1 \cdot x = x$

2. A set X is called a *complex vector space* if $\cdot$ is a mapping $\cdot : \mathbf{C} \times X \to X$ and if the conditions (5-7) in the previous definition hold for all $k, k_1, k_2 \in \mathbf{C}$.

3. An element x of a vector space X is called a *vector*.

In this section we will normally only speak of a vector space, understanding that it can be either a real or a complex vector space. If a result or a definition holds only for either real or complex spaces then we will mention it explicitly. Instead of $k \cdot x$ we will also write $kx = k \cdot x$. In the next definition we introduce the norm of an element in a vector space.

Definition 2.14 Assume X is a vector space.

1. A mapping $\|\| : X \to \mathbf{R}$ is called a *norm* if it satisfies the following conditions:

   - $\|x\| \geq 0$ for all $x \in X$ and $\|x\| = 0$ if and only if $x = 0$,
   - $\|kx\| = |k| \|x\|$ for all constants k and all vectors x
   - $\|x + y\| \leq \|x\| + \|y\|$ for all $x, y \in X$ (Triangle inequality)

2. A *normed space* is a vector space with a norm.

Normed spaces may contain holes and we define therefore complete spaces as spaces that contain all limit points:

Definition 2.15 1. A sequence of vectors $(x_n)_{n \in \mathbf{Z}}$ of a normed space X forms a *Cauchy sequence* if it satisfies the following condition: For all $\epsilon > 0$ there is an index n_0 such that for all $n, m > n_0 : \|x_n - x_m\| < \epsilon$.

2. A normed space X is called *complete* or a *Banach space* if every Cauchy sequence $(x_n)_{n \in \mathbf{Z}}$ converges to an element $x \in X$.

The norm $\|x_n - x_m\|$ measures the distance between the elements x_n and x_m and a Cauchy sequence is thus a sequence whose elements lie eventually arbitrary near each other. One important class of Banach spaces is that of L^p spaces:

Theorem 2.4 [Riesz-Fischer] Define on $L^p(\mathbf{R}^n, \mu)$: $\|f\| = \left(\int |f|^p \, d\mu \right)^{1/p}$. This defines a norm and the spaces $L^p(\mathbf{R}^n, \mu)$ are complete for all $p \geq 1$.

A special type of vector spaces are the vector spaces that possess a scalar product. In these spaces we can measure the length of vectors and angles between vectors:

Definition 2.16 Assume X is a complex vector space. A *scalar product* $\langle \, , \, \rangle$ is a mapping on $X \times X$ which satisfies the following conditions for all $x, y, z \in X$ and all $c \in \mathbf{C}$:

- $\langle \, , \, \rangle : X \times X \to \mathbf{C}$
- $\langle x + y, z \rangle = \langle x, z \rangle + \langle y, z \rangle$
- $\langle cx, y \rangle = c \langle x, y \rangle$
- $\langle x, y \rangle = \overline{\langle y, x \rangle}$
- $\langle x, x \rangle \geq 0$ and $\langle x, x \rangle = 0$ if and only if $x = 0$

($\overline{x}$ is the conjugate complex of x.).

If X is a real vector space then we define a scalar product as a mapping $\langle\,,\,\rangle : X \times X \to$ **R** with the properties given in definition 2.16. Note that we have $\langle x, y \rangle = \langle y, x \rangle$.

Examples are the space $L^2(\mathbf{R}^n, \mu)$ with the scalar product defined as:

$$\langle f, g \rangle = \int fg \, d\mu.$$

and the complex vector space $L^2_{\mathbf{C}}(\mathbf{R}^n, \mu)$ with:

$$\langle f, g \rangle = \int f\overline{g} \, d\mu.$$

On the n-dimensional real and complex spaces $\mathbf{R}^n$ and $\mathbf{C}^n$ we define as usual $\langle x, y \rangle = x'y$ and $\langle x, y \rangle = x'\overline{y}$ where x' is the transposed vector x and $x'y$ is the inner product of the two vectors x and y.

In the spaces $\mathbf{R}^n$ with the common scalar product it is well known that the scalar product of two vectors is proportional to the cosine of the angle between the vectors. We define therefore in the general case:

Definition 2.17 Assume X is a vector space with a scalar product. Then we define:

1. Two elements $x, y \in X$ are said to be *orthogonal* if $\langle x, y \rangle = 0$.

2. Two subsets $M_1, M_2 \subset X$ are said to be *orthogonal* if $\langle x, y \rangle = 0$ for all $x \in M_1$ and all $y \in M_2$.

3. Given that $M \subset X$ is an arbitrary subset of X we define the *orthogonal complement* $M^\perp$ as:
$$M^\perp = \{y \in X \,|\, \langle x, y \rangle = 0 \text{ for all } x \in M\}$$

4. A subset $\{x_i\}$ is called *orthonormal* if $\langle x_i, x_j \rangle = \delta_{ij}$ for all i, j.

For vector spaces with scalar products one gets the following two theorems:

Theorem 2.5 [Cauchy-Schwarz-Inequality] Assume X is a vector space with scalar product $\langle\,,\,\rangle$, then $|\langle x, y \rangle|^2 \le \langle x, x \rangle \langle y, y \rangle$ for all $x, y \in X$.

Theorem 2.6 Assume X is a vector space with scalar product then $\|x\| = \sqrt{\langle x, x \rangle}$ defines a norm on X.

Since a vector space with a scalar product is also a normed space it is meaningful to consider those spaces that are complete in this norm:

Definition 2.18 Assume H is a vector space with scalar product $\langle\,,\,\rangle$ and $\|x\|$ is the norm defined in theorem 2.6. Then we say that H is a *Hilbert space* if H is complete in this norm.

For our purposes the most important examples of Hilbert spaces are the $L^2(\mathbf{R}^n, \mu)$ spaces.

The elements in Hilbert spaces can be decomposed into a series of basis elements which makes the structure of these Hilbert spaces especially simple. We define:

Definition 2.19 Assume H is a Hilbert space.

1. The orthonormal subset $\{x_i\}$ is *complete* or *maximal* if the following condition holds: Assume $\{y_i\}$ is another orthonormal subset of H that contains all vectors x_i. Then the two orthonormal systems $\{x_i\}$ and $\{y_i\}$ are equal as sets.

2. A complete orthonormal subset of H is called a *basis* of H.

For orthonormal subsets of a Hilbert space we have the following inequality:

Theorem 2.7 [Bessel Inequality] Assume $\{x_i\}$ is an orthonormal subset of a Hilbert space H. Then we have the following inequality:

$$\sum \langle x, x_i \rangle^2 \leq \|x\|^2$$

for all $x \in H$.

In the following theorem we collect some conditions which describe the structure of a Hilbert space completely:

Theorem 2.8 Assume H is a Hilbert space and $\{x_i\}$ is an orthonormal subset of H. Then the following statements are equivalent:

- $\{x_i\}$ is complete
- If $\langle x, x_i \rangle = 0$ for all x_i then $x = 0$
- For all $x \in H$ we have the *Fourier decomposition:* $x = \sum \langle x, x_i \rangle x_i$
- For all $x, y \in H$ we have: $\langle x, y \rangle = \sum \langle x, x_i \rangle \langle y, x_i \rangle$.
- For all $x \in H$ we have the *Parseval equality:* $\|x\|^2 = \sum |\langle x, x_i \rangle|^2$.

As in the case of finite-dimensional vector spaces we find therefore that the Hilbert space is the set of all linear combinations of a basis.

We conclude this section on functional analysis with some basic facts about linear functions on Hilbert spaces:

Theorem 2.9 Assume X, Y are two normed spaces and $T : X \to Y$ is a linear operator. Then the we have:

1. The following three conditions are equivalent:
   - T is continuous
   - T is continuous at 0
   - T is continuous at every point $x \in X$.

2. T is continuous if and only if there is a constant $A \geq 0$ such that $\|T(x)\| \leq A\|x\|$ for all $x \in X$.

Theorem 2.10 Assume $A : H_1 \to H_2$ is a continuous, linear function and H_1, H_2 are Hilbert spaces. Then there is exactly one continuous, linear function $A^* : H_2 \to H_1$ such that $\langle Ax, y \rangle = \langle x, A^*y \rangle$ for all $x \in H_1$, $y \in H_2$.

Definition 2.20   1. The operator A^* constructed in theorem 2.10 is called the *adjoint* operator.

2. $A : H \to H$ is called a *hermitian* operator if $A = A^*$, i. e. $\langle Ax, y \rangle = \langle x, Ay \rangle$ for all x, y.

3. If A is a hermitian operator, then we have $\langle Ax, x \rangle \in \mathbf{R}$. A is a *positive* operator if $\langle Ax, x \rangle > 0$ for all $x \in H$.

2.1.5   Exercises

Exercise 2.1 Define **H** as the set of the complex 2×2 matrices of the form

$$\begin{pmatrix} a & b \\ -\bar{b} & \bar{a} \end{pmatrix}$$

This defines the quaternion algebra **H**. Show that:

1. **H** is a two-dimensional complex vector space with basis

$$E_2 = \begin{pmatrix} 1 & 0 \\ 0 & 1 \end{pmatrix}, J = \begin{pmatrix} 0 & 1 \\ -1 & 0 \end{pmatrix}.$$

2. **H** is a four-dimensional real vector space with basis:

$$E_2 = \begin{pmatrix} 1 & 0 \\ 0 & 1 \end{pmatrix}, I = \begin{pmatrix} i & 0 \\ 0 & -i \end{pmatrix}, J = \begin{pmatrix} 0 & 1 \\ -1 & 0 \end{pmatrix}, K = \begin{pmatrix} 0 & i \\ i & 0 \end{pmatrix}.$$

Quaternions are further investigated in the exercises of the next chapter.

Exercise 2.2 Show that all finite subsets of a Hausdorff space are closed.

Exercise 2.3 Define a neighborhood basis as follows: A system of neighborhoods $\mathcal{W} = \{W\}$ is called a *neighborhood basis* if every neighborhood $U(x)$ contains a $W \in \mathcal{W}$: $W \subset U(x)$.
Prove the following statement:
Given a set $\mathcal{W}$ of subsets of $V \subset X$ that satisfies the following conditions:

- $\emptyset \in \mathcal{W}$

- The intersection of any finite number of sets from $\mathcal{W}$ is the union of sets from $\mathcal{W}$.

- The union of all sets $V \in \mathcal{W}$ is X.

then $\mathcal{W}$ is the neighborhood basis of the topology $\mathcal{V}$ obtained by taking all unions of elements in $\mathcal{W}$.

Exercise 2.4 Assume that X is a normed space. Show that the set of all open balls $U(x, r) = \{y \in X \,|\, \|x - y\| < r\}$ where $x \in X$ and $r > 0$ form a neighborhood basis.

Exercise 2.5 [Pythagoras] Assume x and y are orthogonal elements of a Hilbert space H. Show that $\|x + y\|^2 = \|x\|^2 + \|y\|^2$

Exercise 2.6 Assume H is a complex Hilbert space and $A : H \to H$ is a linear map. Show that A is a hermitian operator if and only if $\langle Ax, x \rangle$ is real for all $x \in H$. (Hint: use that if $\langle Ax, x \rangle = 0$ for all $x \in H$ then $A = 0$.)

Exercise 2.7 Use the Cauchy-Schwarz inequality to show that if A is hermitian and $A \geq 0$ then we have for all $x, y \in H$:

$$|\langle Ax, y \rangle|^2 \leq \langle Ax, x \rangle \langle Ay, y \rangle .$$

Exercise 2.8 [Riesz Representation Theorem] Let H be a Hilbert space. For each linear continuous functional $f : H \to \mathbf{C}$ there is a unique element $y_f \in H$ such that: $f(x) = \langle x, y_f \rangle$ for all $x \in H$. (Hint: Define $M = \{x \in H | f(x) = 0\}$ and show that $M^\perp = H/M$ is one-dimensional.)

Exercise 2.9 Assume $\langle \, , \, \rangle$ is the standard scalar product in a real finite dimensional vector space V. Assume further that $f : V \to V$ is a mapping such that $f(x) = Rx$ for a matrix R and $\langle x, x \rangle = \langle f(x), f(x) \rangle$ for all $x \in \mathbf{R}$. Then: $\langle x, y \rangle = \langle f(x), f(y) \rangle$.

A linear mapping that preserves the norm preserves also the angles. (Hint: Consider sums and differences of vectors).

2.2 Algebraic Theory of Groups

In this section we introduce the algebraic concept of a group and we will study some of the basic, algebraic properties of groups. For a more complete treatment the reader is referred to any standard text on algebra, for example [25].

2.2.1 Basic Concepts

Definition 2.21 A non-empty set G is called a *group* if there is a product

$$\circ : G \times G \to G; (g_1, g_2) \mapsto g_1 \circ g_2$$

with the following properties:

- $(g_1 \circ g_2) \circ g_3 = g_1 \circ (g_2 \circ g_3)$. We say that the product is *associative*.
- There is a unique element $e \in G$ such that $e \circ g = g \circ e = g$ for all $g \in G$. The element e is called the *identity element*.
- For all elements $g \in G$ there is a unique element g^{-1} such that $g^{-1} \circ g = g \circ g^{-1} = e$. We call g^{-1} the *inverse* of g.

Note that a group consists of two equally important components, the set G and the composition rule $\circ$. Normally we speak only of the group G, implicitly assuming that it is clear which composition rule is used. We use the notation $(G, \circ)$ only if we want to emphasize a certain composition rule.

Definition 2.22 If the product satisfies the condition

$$g_1 \circ g_2 = g_2 \circ g_1$$

for all $g_1, g_2 \in G$ then we say that G is a *commutative group* or an *abelian group*.

In the general case we often write $g_1 g_2$ instead of $g_1 \circ g_2$ and we denote the identity by 1. In the commutative case we write $g_1 + g_2$ instead of $g_1 \circ g_2$. In this case we denote the identity element by 0 and the inverse element of g by $-g$.

Definition 2.23   1. If G has a finite number of elements then we say that G is a *finite group;* otherwise we say that G is *infinite.*

2. The number of elements in the group is called the *order* of the group and it is denoted by $|G|$.

The simplest example of a group is a set with a single element e and the product $e \circ e = e$. The group with two elements e and a is obviously given by defining the product as: $e \circ e = e; e \circ a = a \circ e = a$ and $a \circ a = e$. Often it is convenient to describe the product rule with the help of a table. In the previous case this table is given by:

$$
\begin{array}{c|cc}
 & e & a \\
\hline
e & e & a \\
a & a & e
\end{array}
$$

Given a set G of three elements e, a and b we can make G into a group by defining the group multiplication $\circ$ with the following table:

$$
\begin{array}{c|ccc}
 & e & a & b \\
\hline
e & e & a & b \\
a & a & b & e \\
b & b & e & a
\end{array}
\tag{2.4}
$$

Up to now we used G as a mere set of symbols. In our applications these symbols will always represent objects. One way to give the symbols in the last group a meaning is to consider the following mapping:

$$e \mapsto 0; a \mapsto 1; b \mapsto 2.$$

The multiplication table 2.4 now becomes:

$$
\begin{array}{c|ccc}
 & 0 & 1 & 2 \\
\hline
0 & 0 & 1 & 2 \\
1 & 1 & 2 & 0 \\
2 & 2 & 0 & 1
\end{array}
\tag{2.5}
$$

The group operation is addition modulo 3 in this interpretation of the group.

Instead of interpreting the group elements as numbers we could also give them a geometrical meaning. For this purpose we define the following map:

$$e \mapsto E; a \mapsto R; b \mapsto R^2$$

where E is the 2×2 identity matrix, R is the matrix

$$R = \begin{pmatrix} -1/2 & -\sqrt{3}/2 \\ \sqrt{3}/2 & -1/2 \end{pmatrix}$$

of 120 degree rotation and R^2 is the matrix of 240 degree rotation. In this case the group operation becomes the concatenation of two rotations or, equivalently the product of two matrices.

A slightly different geometric interpretation of G is the following: Consider set X consisting of the three points x, y and z with coordinates: $x = (0, 1/\sqrt{3})$; $y = (-1/2, 1/(2\sqrt{3}))$ and $z = (1/2, 1/(2\sqrt{3}))$. As group elements we use the three functions $e, a, b : X \to X$ defined as:

$$e(p) = p \text{ for all } p \in X$$

$$a(x) = y; \quad a(y) = z; \quad a(z) = x$$
$$b(x) = z; \quad b(y) = x; \quad b(z) = y$$

Geometrically e, a and b act again as rotations but we can also interpret them as permutations acting on the set X. The situation where X is an ordinary set and G is a group of functions $f : X \to X$ will be treated extensively in the following.

The following examples introduce groups that will be studied in detail in the following chapters.

Examples 2.2   1. **The additive group of real numbers:**   The set of real numbers becomes a group under the usual addition: $x_1 \circ x_2 = x_1 + x_2$. The identity is the number zero and the inverse of x is $-x$. This group will be denoted by **R**.

2. **The additive group of integers:**   The set of integers under the usual addition is a group. It is denoted by **Z**.

3. **The multiplicative group of non-zero real numbers:**   The set of the real numbers without the number zero is a group under the usual multiplication. The identity is the number one and the inverse element of x is $x^{-1} = 1/x$. This group is denoted by $\mathbf{R}_0$.

4. **The multiplicative group of positive real numbers:**   The set of the positive real numbers is a group under the usual multiplication. This group is denoted by $\mathbf{R}_+$.

5. **The Symmetrical Group:**   Assume X is a set consisting of N elements. A bijective map $p : X \to X$ is called a *permutation* of X. The permutations of X form a group under the usual composition of functions. This group is called the *symmetrical group* and it is denoted by S_N.

6. **The n-dimensional space:** The set of all n-dimensional vectors with real components is a group under the normal addition of vectors. This group will be denoted by $\mathbf{R}^n$. Note that also the set of vectors (without the group operation) is denoted by $\mathbf{R}^n$.

7. **Rotations in 2-D:**   The rotation R_ϕ is defined as the mapping $R_\phi : \mathbf{R}^2 \to \mathbf{R}^2$; $(x, y) \mapsto (x', y')$ where x' and y' are defined as:

$$x' = x \cos \phi - y \sin \phi \qquad (2.6)$$
$$y' = x \sin \phi + y \cos \phi \qquad (2.7)$$

The product $R_{\phi_2} \circ R_{\phi_1}$ is the rotation obtained by first performing the rotation R_{ϕ_1} and then the rotation R_{ϕ_2}. We find that $R_{\phi_1 + \phi_2} = R_{\phi_1} \circ R_{\phi_2}$. It is easy to see that these rotations define a group under the usual composition of mappings. This group is denoted by $SO(2)$.

8. **Complex Numbers:**   The complex numbers under addition and the non-zero complex numbers under multiplication form groups. These groups are denoted by $\mathbf{C}$ and $\mathbf{C}_0$ respectively. The unit circle, i.e. the set $C = \{z : |z| = 1\}$ forms a group under complex multiplication.

9. **The open interval $[0, a)$:** The open interval is a group under addition modulo a.

10. **The general linear group:**   The set of complex matrices of size $n \times n$ with non-zero determinant is a group under ordinary matrix multiplication. It is denoted by $GL(n, \mathbf{C})$. In the same way we define $GL(n, \mathbf{R})$ as the set of real $n \times n$ matrices with non-zero determinant.

11. **The special linear groups:**   The sets of real or complex matrices of size $n \times n$ with determinant 1 are groups. These groups are denoted by $SL(n, \mathbf{R})$ and $SL(n, \mathbf{C})$ respectively.

12. **The orthogonal groups:**   An *orthogonal* matrix R is an element of $GL(n, \mathbf{R})$ that satisfies the condition: $R'R = E_n$ where R' is the transpose of R and E_n is the $n \times n$ identity matrix. The set of all orthogonal matrices defines a group under matrix multiplication. It is denoted by $O(n)$.

13. **The special orthogonal groups:**   The set of all orthogonal matrices $R \in O(n)$ with $\det R = 1$ is a group, the special orthogonal group $SO(n)$. Note that we denoted both, the matrix group and the group of 2-D rotations by $SO(2)$. This reflects the known property that rotations can be described by matrices from $SO(n)$ after the introduction of an appropriate coordinate system. We could also view these two groups as realizations of the same abstract group.

14. **The unitary groups:** Recall that a complex matrix U is *unitary* if $U^*U = E_n$ where U^* is the transposed, conjugate complex of U. The set of all $n \times n$ unitary matrices forms a group under matrix multiplication. This group is denoted by $U(n)$.

15. **The special unitary groups:**   The set of all unitary matrices of size $n \times n$ with $\det U = 1$ is also a group, the special unitary group. These groups are denoted by $SU(n)$.

16. **Euclidian motion groups:**   Recall that a euclidian motion is a rotation followed by a translation. The euclidian motions in n-dimensional space form a group that is denoted by $M(n)$. After introducing coordinates we can describe a euclidian motion as $y = Rx + T$ where x and y are the coordinate vectors of the original point and its image. Here $R \in SO(n)$ is a rotation matrix and T is a translation vector. If we introduce the $(n+1)$–dimensional vector $\tilde{x}$ as $\tilde{x} = \begin{pmatrix} x & 1 \end{pmatrix}$ then we can also write $\tilde{y} = M\tilde{x}$ where M is the matrix: $M = \begin{pmatrix} R & T \\ 0 & 1 \end{pmatrix}$.

2.2.2   Subgroups and Cosets

Definition 2.24   1. A non-empty subset H of a group G is called a *subgroup* if $g_1 g_2^{-1} \in H$ for all g_1 and g_2 in H.

2. If $H \neq G$ then we say that H is a *proper* subgroup of G.

We find $e = g_1 g_1^{-1} \in H$ and therefore also $g^{-1} = eg^{-1} \in H$. Thus H is itself a group under the composition rule inherited from the larger group G. If H is a subgroup of G then we write $H < G$.

Examples 2.3   1. The subset $H = \{e\}$ is the smallest subgroup of a group. G is its largest subgroup. G and $\{e\}$ are the *trivial* subgroups of G.

2. The intersection of two subgroups is a subgroup.

3. Assume a is a fixed element of G and define $H = \{a^n | n \in \mathbf{Z}\}$. Then H is a subgroup of G. Such subgroups are called *cyclic*. The element a is called the *generator* of H.

4. The set of 3-D rotations around the z-axis forms a subgroup of $SO(3)$.

We have the following relations between the groups introduced so far:

Theorem 2.11   1. $\mathbf{Z} < \mathbf{R}$

2. $\mathbf{R}_+ < \mathbf{R}_0$

3. $\mathcal{C} < \mathbf{C}_0$

4. $SU(n) < U(n) < GL(n, \mathbf{C})$

5. $SO(n) < O(n) < GL(n, \mathbf{R})$

6. $SO(n) < SU(n)$

7. $O(n) < U(n)$

8. $GL(n, \mathbf{R}) < GL(n, \mathbf{C})$

Definition 2.25   1. Assume H is a subset of G and g is an arbitrary element of G, then we denote by Hg the set $\{hg | h \in H\}$. Such a set is called a *right coset* of H. *Left cosets* are defined in like manner.

2. An element of a coset is called a *representative* of the coset and by $\tilde{g}$ we denote the coset which contains g.

Theorem 2.12 The group G falls into pairwise disjoint cosets of H if $H < G$.

It is clear that G is the union of all its cosets and it only remains to show that different cosets are disjoint. To see this assume that g is a common member of the cosets Hg_1 and Hg_2. Then we find that $g = h_1 g_1 = h_2 g_2$ for some elements $h_1, h_2 \in H$. If hg_1 is an arbitrary element in Hg_1 then we find $hg_1 = hh_1^{-1} h_1 g_1 = hh_1^{-1} h_2 g_2 = h' g_2$ with $h' \in H$ and therefore we find that hg_1 is an element of Hg_2 which shows that the two cosets are equal.

Definition 2.26 Assume K is a subset of the group G. Then we say that K generates G if all elements $g \in G$ have the form: $g = a_1 ... a_n$ where $a_i \in K$ or $a_i^{-1} \in K$. In general the index n is different for different elements $g \in G$.

Thus a subset K generates G if all elements in G are products of elements in K or their inverses.

Examples 2.4 The group of permutations of three elements S_3 has six elements:

$$e = \begin{pmatrix} 1 & 2 & 3 \\ 1 & 2 & 3 \end{pmatrix} \ a = \begin{pmatrix} 1 & 2 & 3 \\ 2 & 3 & 1 \end{pmatrix} \ b = \begin{pmatrix} 1 & 2 & 3 \\ 3 & 1 & 2 \end{pmatrix}$$

$$c = \begin{pmatrix} 1 & 2 & 3 \\ 1 & 3 & 2 \end{pmatrix} \ d = \begin{pmatrix} 1 & 2 & 3 \\ 3 & 2 & 1 \end{pmatrix} \ f = \begin{pmatrix} 1 & 2 & 3 \\ 2 & 1 & 3 \end{pmatrix}.$$

The group operation is summarized in the following table:

	e	a	b	c	d	f
e	e	a	b	c	d	f
a	a	b	e	f	c	d
b	b	e	a	d	f	c
c	c	d	f	e	a	b
d	d	f	c	b	e	a
f	f	c	d	a	b	e

(2.8)

The group is not commutative since $ac \neq ca$. The set $H = \{e, a, b\}$ is a cyclic subgroup and the subset $K = \{a, c\}$ generates S_3.

2.2.3 Normal Subgroup and Center

Definition 2.27   1. Let H be a subgroup of the group G. Consider each right coset Hg as a single element in the space of right cosets. The new space of cosets is called the *factor space* or the *space of cosets* of the subgroup H in G. It is denoted by G/H.

2. If the number of elements in G/H is finite then we call this number the *index of H in G* and denote it by $|G/H|$.

For finite groups and their subgroups we have the following relations:

Theorem 2.13 Let G be a finite group and $H < G$ be a subgroup. Then we have:

1. $|G| = |H| \, |G/H|$.

2. The order of H divides the order of G.

3. G has only trivial subgroups if the order of G is a prime number.

Examples 2.5   1. Take $G = \mathbf{R}$ and $H = \mathbf{Z}$. A representative $\tilde{g} \in G/H$ has the form: $\tilde{g} = g_\alpha = \{\alpha + n | n \in \mathbf{Z}, 0 \leq \alpha < 1\}$.

2. If $G = \mathbf{Z}$ and H is the set of all even integers then $|G/H| = 2$ and we can think of the elements of G/H as the sets of even and odd integers.

Normally left cosets and right cosets are different. The subgroups for which left cosets and right cosets are identical are called normal:

Definition 2.28   1. Let G be a group and $H < G$. H is called a *normal divisor* or a *normal subgroup* of G if $gH = Hg$ for all $g \in G$.

    2. $\{e\}$ and G are called the *trivial normal divisors* of G. All other normal subgroups are called *nontrivial*.

    3. A group is called *simple* if it has only trivial normal subgroups.

Theorem 2.14 Assume H is a normal subgroup of G. On the factor space G/H we define the multiplication: $g_1 H g_2 H = (g_1 g_2) H$. This definition is independent of the elements g_1 and g_2 and G/H becomes a group under this operation.

Definition 2.29 The set G/H together with the multiplication defined in theorem 2.14 is called the *factor group G/H*.

Examples 2.6   1. $G = GL(n, \mathbf{C})$, $H = SL(n, \mathbf{C})$. H is a normal subgroup of G. The map $\lambda : G/H \to \mathbf{C}_0; \tilde{g} \mapsto \det g$ is bijective and satisfies $\lambda(\tilde{g}_1 \tilde{g}_2) = \lambda(\tilde{g}_1)\lambda(\tilde{g}_2)$ for all $\tilde{g}_1, \tilde{g}_2 \in G/H$. An element in G/H has the form $\tilde{g}_\alpha = \{g \in GL(n, \mathbf{C}) | \det g = \alpha\}$.

    2. $G = \mathbf{R}$, $H = \mathbf{Z}$, $G/H = \{g_\alpha | \alpha \in [0, 1)\}$ with $g_\alpha = \{\alpha + n | n \in \mathbf{Z}\}$. The addition in G/H becomes: $g_{\alpha_1} + g_{\alpha_2} = g_{[\alpha_1 + \alpha_2]}$ where we defined: $[\alpha_1 + \alpha_2] = \alpha_1 + \alpha_2 \bmod 1$.

Finally we introduce a subgroup that describes "how commutative" a group is.

Definition 2.30 The *center* of a group is the set of all group elements that commute with all group elements, it will be denoted by $Z(G)$.
$Z(G) = \{g_0 \in G | g_0 g = g g_0 \text{ for all } g \in G\}$.

Theorem 2.15 The center $Z(G)$ of a group has the following properties:

    1. $Z(G)$ is a group.

    2. $Z(G)$ is commutative.

    3. $Z(G)$ is a normal subgroup of G.

2.2.4   Homomorphisms and Isomorphisms

In this section we will discuss mappings between groups which preserve the group operation.

Definition 2.31   1. A mapping $f : G \to G'$ of a group G into a group G' is called a *homomorphism* if it preserves the group operation, i. e.: $f(g_1 g_2) = f(g_1)f(g_2)$ for all $g_1, g_2 \in G$.

    2. A bijective homomorphism is called an *isomorphism*.

3. Two groups G and G' are called *isomorphic* if there is an isomorphism $f : G \to G'$. In this case we write $G \cong G'$.

Two isomorphic groups are thus identical as far as their algebraic properties are concerned. Examples are the group of 2-D rotations, the group of all complex numbers of magnitude one and the group of all real, orthogonal 2×2 matrices.

Given two groups G and G' and a homomorphism f between them it is easy to show that the set of all elements in G that are mapped to the identity element in G' forms a subgroup of G. We define:

Definition 2.32 $\ker(f) = \{g \in G | f(g) = e'\}$ is called the *kernel* of f. e' is in this case the identity in G'.

Homomorphisms have the following properties:

Theorem 2.16 A homomorphism f has the following properties:

1. $f(e) = e'$
2. $f(g^{-1}) = f(g)^{-1}$
3. $\ker(f)$ is a normal subgroup of G.
4. f is an isomorphism if f is surjective and if $\ker(f) = \{e\}$.

The case where an isomorphism maps a group onto itself is so important that we consider this case separately:

Definition 2.33   1. An isomorphism of G onto itself is called an *automorphism*.

2. The automorphisms form a group under the usual composition rule $(f_2 f_1)(g) = f_2(f_1(g))$, this group is called the *automorphism group* of G and it is denoted by $A(G)$.

3. An automorphism of the form: $f(g) = g_0^{-1} g g_0$ for a fixed element $g_0 \in G$ is called an *inner automorphism*.

4. The inner automorphisms form a subgroup of $A(G)$, the *inner automorphism group* of G. It is denoted by $A_i(G)$.

An important example of a homomorphism is the projection from the original group onto a factor group. We define:

Definition 2.34 Assume that G is a group and H is a normal subgroup of G and G/H is the factor group. The mapping: $p : G \to G/H; g \mapsto Hg$ is called the *canonical homomorphism* or the *natural mapping*.

We saw that the kernel of a homomorphism is a normal subgroup and we can therefore construct the factor group $G/\ker f$. For this factor group we find the following important theorem which decomposes an arbitrary homomorphism into the projection and an isomorphism.

Theorem 2.17 Assume $f : G \to G'$ is a surjective homomorphism and $H = \ker(f)$. Then

1. $G' \cong G/H$

2. $f = gp$ where g is an isomorphism and p is the canonical homomorphism (see Figure 2.1).

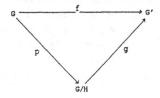

Figure 2.1: Factorization of a homomorphism

2.2.5   Transformation Groups

In this section we will generalize the example from page 17 where we considered permutations and rotations that left a triangle invariant. We considered the set of the three corner points of a triangle and a set of rotations that mapped a given corner point into another corner point. The same concept underlies the definition of the symmetrical group (see page 17 where we had a set of N elements and the group elements of S_N where the bijective mappings of this set. We generalize these examples in the concept of a transformation group.

Definition 2.35 Let X be any set.

1. A bijective map of X is called a *transformation* of X.

2. If g is a transformation then we write xg or x^g instead of $g(x)$ and we say that g is a *right transformation*. If we write gx or $^g x$ instead of $g(x)$ then we say that g is a *left transformation*.

3. Assume g_1 and g_2 are right transformations, then we define the *product of g_1 and g_2* as the transformation which is obtained by applying first g_1 and then g_2. We write $x(g_1 g_2) = (x g_1)g_2$.

4. The set of all transformations forms a group under the product defined above. This group is denoted by $G(X)$.

5. Any subgroup of $G(X)$ is called a *transformation group of the set X*.

6. The pair (X, G) with a set X and a transformation group G is called a *space X with transformation group G*.

7. The subgroup of all linear transformations of X is denoted by $GL(X)$.

Examples 2.7   1. The symmetric group S_N is the special transformation group that belongs to a finite set with N elements.

2. Assume C_r is a circle with radius r around the origin and $SO(2)$ is the group of 2-D rotations. The pair $(C_r, SO(2))$ is a space with a transformation group. It is also easy to see that the unit sphere and the group $SO(n)$ form a transformation group.

3. Let $\mathbf{R}$ be the real axis and S the group of transformations defined by $sx = ax + b$ with $a \neq 0$. S is called the *group of linear transformations of the line*. The subgroup S^1 of all transformations of the form $sx = x + b$ is called the *group of translations of the line* and the subgroup S^2 defined by $sx = ax$ is the *group of dilatations of the line*. $(\mathbf{R}, S), (\mathbf{R}, S^1)$ and $(\mathbf{R}, S^2)$ are transformation groups.

4. Let Π^1 be the complex plane together with an additional point ∞. Let F^1 be the set of transformations $f : z \mapsto z' = f(z)$ defined as:

$$z' = \begin{cases} \frac{az+b}{cz+d} & \text{if } z \neq \infty \text{ and } z \neq -d/c \\ \infty & \text{if } z = -d/c \\ \frac{a}{c} & \text{if } z = \infty \end{cases}$$

where a, b, c, d are arbitrary complex numbers with $ad - bc \neq 0$. Then (Π^1, F^1) form a space with a transformation group and F^1 is called the *group of fractional-linear transformations of the plane Π^1*.

Two points in the space belonging to a transformation group can be regarded as similar if there is a transformation that maps one point onto the other. For example all points on a circle with a certain radius are similar with respect to the rotation group. We formalize this in the following definition:

Definition 2.36 Assume (X, G) is a space X with transformation group G, assume further that x is an element in X. Then we call the set $\{xg | g \in G\}$ an *orbit* or a *trajectory* relative to G.

It is easy to see that "being in the same orbit" defines an equivalence relation and X is therefore the disjoint union of orbits.

Definition 2.37 The space X is called *transitive* or *homogeneous* if X consists of one orbit.

Homogeneous spaces are important in our applications since we can identify points in the underlying space X with equivalence classes of group elements in the following way:

Assume x_0 is an arbitrary but fixed point in the space X and define the subset $H \subset G$ as the set of all elements that leave x_0 fixed:

$$H = \{g \in G : x_0 g = x_0\}$$

For an element $g \in G$ we define the subset H^g as $g^{-1}Hg$ and K as $K = \{H^g : g \in G\}$. On G we define an equivalence relation by $g_1 \equiv g_2$ if they are elements of the same set H^g. This means that there is a $g \in G$ and elements $h_i \in H$ such that $g_i = g^{-1}h_ig (i = 1, 2)$. This is obviously an equivalence relation in G and G is thus the disjunct union of the different elements in K.

Now assume that $y \in X$. Then we can find an element $g_y \in G$ such that $y = x_0 g_y$ since X is homogeneous. We define the mapping i as:

$$i : X \to K; y \mapsto H^{g_y}$$

This map is well defined since $y = x_0 g = x_0 g'$ implies $g'g^{-1} \in H$ and therefore $H^g = H^{g'}$ and it is also bijective. Another way to map X into the group G is to identify a point $y \in$

X with all transformations $g \in G$ that map x_0 into y. Obviously these transformations are the elements of the right coset Hg_y where g_y is defined as above. Consequently we can identify the space X (on which G operates) with the factor space of all right cosets of H.

In one of the simplest examples X is the unit circle and $G = SO(2)$. In this case we identify X with $SO(2)$ by mapping the complex number $e^{i\phi}$ to the rotation with rotation angle ϕ. If X is the unit sphere in 3-D space and if we select x_0 as the north pole then we find that H is the set of all 3-D rotations that leave the north pole fixed. The elements in H are thus represented by matrices of the form $\begin{pmatrix} R & 0 \\ 0 & 1 \end{pmatrix}$ where $R \in SO(2)$ is a 2-D rotation.

One important example of a transformation group can be obtained by using as X the group G itself. This example is formulated in the next definition and theorem (see also definition 2.33)

Definition 2.38 1. Two elements g_1 and g_2 in G are called *conjugate* if there is an element $g \in G$ such that $g_2 = g^{-1}g_1g$

2. The set of all elements in G which are conjugate to one fixed element $g \in G$ is called a *conjugacy class*.

From the definition of an inner automorphism and the definition of a conjugacy class we obtain immediately the following theorem:

Theorem 2.18 1. The orbit of an element g relative to the inner automorphism group $A_i(G)$ is the conjugacy class of g.

2. Define the homomorphism $f : G \to A_i(G); g \mapsto a_g$ where a_g is the inner automorphism $a_g(h) = g^{-1}hg$. Then we find that $\ker(f) = Z(G)$.

3. A subgroup $H < G$ is a normal subgroup if and only if H is invariant under all inner automorphisms $a_g, g \in G$.

2.2.6 Direct Product of Groups

The last concept from the elementary, algebraic theory of groups is the direct product of a finite number of groups which will be introduced in this section.

Definition 2.39 Assume $G_1, ..., G_n$ is a set of groups. Then $G_1 \times ... \times G_n$ is the set of all sequences $\{(g_1, ..., g_n)|g_i \in G_i\}$. We make this set product into group by defining a composition rule $\circ$ as $(g_1, ..., g_n) \circ (g_1', ..., g_n') = (g_1 g_1', ..., g_n g_n')$ where $g_i g_i'$ is the group operation defined on the group G_i. We will also write $(g_1, ..., g_n)(g_1', ..., g_n')$ instead of $(g_1, ..., g_n) \circ (g_1', ..., g_n')$. The cartesian product of the groups G_i together with this composition rule is called the *direct product of the groups*. The identity element is given by $(e_1, ..., e_n)$ where e_i is the identity element in G_i.

Theorem 2.19 If we define the mappings
$f_i : G_i \to G_1 \times ... \times G_n; g_i \mapsto (e_1, ..., g_i, ..., e_n)$ then we can identify G_i with a subgroup of the product $G = G_1 \times ... \times G_n$ and we write g_i instead of $(e_1, ..., g_i, ..., e_n)$. With these notations it is easy to show that all elements $g \in G$ are of the form: $g = g_1...g_n$ and that any two elements $g_j, g_k \in G$ with $j \neq k$ satisfy $g_j g_k = g_k g_j$.

Definition 2.40 Assume $G_1, ..., G_n$ are subgroups of G which satisfy the following conditions: All elements $g \in G$ are of the form $g = g_1...g_n$ with $g_i \in G_i$ and they also satisfy the condition $g_j g_k = g_k g_j$ if $j \neq k$. Then we say that G is the *direct product* of the subgroups $G_1...G_n$.

2.2.7 Exercises

Exercise 2.10 Show the group properties for the following groups:

1. **A finite group:** The set $G_0 = \{1, i, -1, -i\}$ is a finite group under the usual complex multiplication.

2. **Multiples of a fixed natural number:** The set is given by $\{m = np | n \in \mathbf{Z}\}$, the composition is the addition of natural numbers. This group is denoted by $\mathbf{Z}_p$.

3. **The pth roots of unity:** This group consists of the numbers $e^{2\pi i k/p}$ and the composition rule is complex multiplication (k is an integer and p is a natural number). This group is denoted by Ω_p.

4. **A dynamical system:** Assume the function $f(t)$ describes the state of a system at time t, assume further that t_0 is a fixed constant and define the operator T as $(Tf)(t) = f(t + t_0)$. Show that the operators $\{T^n | n \in \mathbf{Z}\}$ form a group under the composition rule $T^n T^m = T^{n+m}$.

Exercise 2.11   1. Show that $SL(n, \mathbf{C})$ is a normal subgroup of $GL(n, \mathbf{C})$.

2. Show that $GL(n, \mathbf{C})/SL(n, \mathbf{C})$ and $\mathbf{C}_0$ (i.e. the multiplicative group of non-zero complex numbers) are isomorphic. What is the isomorphism?

3. Are $\mathbf{R}_0$ and $GL(n, \mathbf{R})/SL(n, \mathbf{R})$ isomorphic?

Exercise 2.12 What is $SO(3)/SO(2)$?
Hint: A 3-D rotation can be described by its rotation axis and its rotation angle.

Exercise 2.13 Prove theorem 2.15.

Exercise 2.14 Show that the center $Z(GL(n, \mathbf{C})$ is given by the matrices λE_n where $\lambda \in \mathbf{C}$ and E_n is the unit matrix in G. (Hint: Find all matrices that commute with diagonal matrices that have n different non-zero entries in the diagonal.) What is $Z(GL(n, \mathbf{R}))$?

Exercise 2.15 Prove the properties of a homomorphism mentioned in theorem 2.16.

Exercise 2.16 Show that $f : GL(n, \mathbf{C}) \to \mathbf{C}; g \mapsto \det g$ is a homomorphism.

Exercise 2.17 Show that every group G is isomorphic to the group of all right translations on G.

Exercise 2.18 $[0, 1)$ together with the addition modulo 1 is isomorphic to the unit circle $\{e^{2\pi i \phi}\}$ with complex multiplication.

Exercise 2.19 $[0,1)$ together with addition modulo 1 is isomorphic to $SO(2,\mathbf{R})$.

Exercise 2.20 Assume G is a finite group with p elements, $|G| = p$ with a prime number p. What are the cyclic subgroups of G?

Exercise 2.21 Define **H** as the set of the complex 2×2 matrices of the form

$$\begin{pmatrix} a & b \\ -\bar{b} & \bar{a} \end{pmatrix}$$

1. Show that this set is a group with matrix addition.

2. Show that the non-zero elements in **H** form a group under matrix multiplication. Is this group commutative?

3. Show that the matrices

$$\begin{pmatrix} a & 0 \\ 0 & a \end{pmatrix}$$

   with $a \in \mathbf{R}$ form a subgroup (under addition and multiplication).

4. Show that **R** is the center of **H** under multiplication.

5. Show that the matrices

$$\begin{pmatrix} c & 0 \\ 0 & \bar{c} \end{pmatrix}$$

   with $c \in \mathbf{C}$ form a subgroup (under addition and multiplication).

6. Show that

$$\mathbf{C}^2 \to H; (a,b) \mapsto \begin{pmatrix} a & b \\ -\bar{b} & \bar{a} \end{pmatrix}$$

   is an isomorphism.

The previous sequence of exercises shows that we have the following sequence of extensions of $\mathbf{R}$: $\mathbf{R} \subset \mathbf{C} \subset \mathbf{H}$. The set **H** together with addition and multiplication is called the *quaternion algebra* **H**. The theorem of Frobenius shows that **C** and **H** are essentially the only extensions of **R**. For a proof see [40]. More information on quaternions, their relation to other number systems and a historical background can be found in [18].

2.3 Topological Groups

In this section we combine the algebraic concept of a group and the topological concept of continuity to define topological groups, i.e. groups in which the group operations are continuous maps. We will also list some of the basic properties of these groups and we will study some matrix groups that will play a central role in the following chapters.

2.3.1 Topological Groups

We first introduce some notation: Assume $U \subset G$ is a subset of G. Then we define the following subsets:

$$U^{-1} = \left\{ g^{-1} | g \in U \right\},$$

$$g_0 U = \{ g_0 g | g \in U \},$$

and if U_1 and U_2 are subsets of G then we define $U_1 U_2$ as the subset

$$U_1 U_2 = \{ gh | g \in U_1, h \in U_2 \} .$$

We now define a topological group as a group with continuous group operations:

Definition 2.41 A group G is called a *topological group* if:

- G is a Hausdorff space and if
- the group operation $\circ : G \times G \to G; (g, h) \mapsto g \circ h$ and the inversion $G \to G; g \mapsto g^{-1}$ are continuous maps.

Theorem 2.20   1. The mapping $f_i : g \mapsto f_i(g) = g^{-1}$ is a homeomorphism.

2. The left translations $g \mapsto g_0 g$ and the right translations $g \mapsto g g_0$ are homeomorphisms.

To see the first part note that f_i is continuous according to the definition and that $f_i^{-1} = f_i$. In order to show the second part we use the fact that $g \mapsto g g_0^{-1}$ is the inverse of the translation $g \mapsto g g_0$.

Examples 2.8   1. An ordinary group equipped with the discrete topology is a topological group.

2. The real and complex spaces $\mathbf{R}^n$ and $\mathbf{C}^n$ are additive, topological groups under their natural topologies.

3. The non-zero real and the complex numbers $\mathbf{R}_0$ and $\mathbf{C}_0$ are multiplicative, topological groups under their natural topologies.

4. Consider $GL(n, \mathbf{R})$ and $GL(n, \mathbf{C})$ as subspaces of $\mathbf{R}^{n^2}$ and $\mathbf{C}^{n^2}$ respectively. As subsets they inherit the topology from the larger spaces. This topology is called the **natural topology** of these groups. These groups form topological groups since polynomials are continuous functions.

We continue by defining topologies on subgroups and factor groups of topological groups:

Definition 2.42 Assume H is a subgroup of a topological group G. Then we make H into a topological group by providing H with the topology of a subspace. An open set V in H is thus a subset of the form $V = U \bigcap H$ where U is an open subset of G.

The factor group is given a topology by the following construction:

Definition 2.43 Assume G is a topological group and H is a subgroup of G. G/H is the set of right cosets and $p : G \to G/H; g \mapsto Hg$ is the natural or canonical mapping. Then we define a topology on G/H as the set of all images of open sets $U \subset G$ under p. A subset $V \subset G/H$ is therefore open if there is an open set $U \subset G$ with $V = p(U)$.

Theorem 2.21   1. We say that a map is open if the image of every open set is open. The natural mapping p is thus open and continuous.

2. If H is a closed subgroup of G then G/H is separated.

3. If H is a closed, normal subgroup of G then G/H is a topological group.

Next we have to define the mappings that are compatible with the topological and algebraical properties of groups:

Definition 2.44 Assume G and G' are topological groups and $f \to G'$ is a mapping. Then we say that

1. f is a *continuous homomorphism* if f is a homomorphism and if it is continuous.

2. f is an *continuous isomorphism* if f is an isomorphism and if it is continuous.

3. f is a *topological isomorphism* if f is an isomorphism and if it is a homeomorphism.

1. Two groups are called *topological isomorphic* if there is a topological isomorphism between them.

2. A topological isomorphism $f : G \to G$ is called a *topological automorphism*.

The theorem 2.17 about the factorization of a homomorphism now becomes:

Theorem 2.22   1. If $f : G \to G'$ is a continuous surjective homomorphism and $H = \ker(f)$, then:

- H is a closed, normal subgroup of G.

- $f = gp$ where g is a continuous isomorphism of G/H onto G'.

2. If f is also open then g is a topological isomorphism.

In the last definition of this section we define continuous vector-valued functions and continuous linear operators on the group.

Definition 2.45   1. Assume $f : G \to \mathbf{C}^n; g \mapsto (f_1(g), \ldots, f_n(g))$ is a complex valued vector function then we say that f is a *continuous function* if all functions f_i are continuous.

2. Denote the space of all linear operators on a vector space X by $L(X)$. If X is finite dimensional then the elements of $L(X)$ are described by matrices. Now assume that $A : G \to L(X); g \mapsto A(g)$ is an operator valued function. For each $g \in G$ the function value $A(g)$ is a linear operator on a finite dimensional vector space X. Then we say that A is a *continuous operator* if the mapping $g \mapsto A(g)x$ is continuous for all $x \in X$.

It is easy to see that a linear operator is continuous if and only if its matrix elements $(a_{ij})(g)$ are continuous, complex-valued functions in any basis $e_1, \ldots, e_n$ of X.

2.3.2   Some Matrix Groups

In this section we derive some important properties of the matrix groups that will be most important in what follows.

In section 2.2 we introduced the following matrix groups: $GL(n, \mathbf{R})$, $GL(n, \mathbf{C})$, (the general linear groups), $SL(n, \mathbf{R})$, $SL(n, \mathbf{C})$, (the special linear groups), $O(n)$, $SO(n)$, (the orthogonal and special orthogonal groups) and $U(n)$ and $SU(n)$ (the unitary and special unitary groups). There we mentioned that these sets form groups under the usual matrix multiplication. From the definition of these groups it is also easy to see that we have the following subgroup relations among these groups (see also theorem 2.11):

1. $SO(n) < O(n)$,

2. $SO(n) < SL(n, \mathbf{R}) < GL(n, \mathbf{R})$,

3. $SU(n) < U(n)$ and

4. $SU(n) < SL(n, \mathbf{C}) < GL(n, \mathbf{C})$.

We also mentioned (see section 2.11) that $SL(n, \mathbf{R})$ and $SL(n, \mathbf{C})$ are normal subgroups of the general linear groups $GL(n, \mathbf{R})$ and $GL(n, \mathbf{C})$ respectively. Furthermore we saw that $\mathbf{C_0}$ and $\mathbf{R_0}$ (the non-zero complex and real numbers) were isomorphic to the factor groups $GL(n, \mathbf{C})/SL(n, \mathbf{C})$ and $GL(n, \mathbf{R})/SL(n, \mathbf{R})$ respectively. It can also be shown that the mapping

$$f : GL(n, \mathbf{C}) \to \mathbf{C_0}; g \mapsto \det g$$

is an open, continuous isomorphism. The groups $\mathbf{C_0}$ and $GL(n, \mathbf{C})/SL(n, \mathbf{C})$ are thus topologically isomorphic. In this section we will derive some important properties of these groups.

Theorem 2.23 $U(n), SU(n), O(n)$ and $SO(n)$ are compact groups.

We consider only the unitary case since the compactness of the other groups can be shown by similar arguments:

From elementary analysis it is known that a subset of $\mathbf{R}^n$ is compact if and only if it is closed and bounded. From the orthogonality relations $\sum_{l=1}^{n} u_{lk}\overline{u_{lj}} = \delta_{ij}$ we find that $U(n)$ is closed since the left side of the equation is a continuous function in the variables u_{kl} and the right side is a closed set. From the orthogonality we find also $\sum_{k=1}^{n}\sum_{l=1}^{n} |u_{lk}|^2 = n$ and the matrices in $U(n)$ are thus all contained in a sphere of radius n. $U(n)$ is thus closed and bounded and therefore compact. $SU(n) = U(n) \bigcap SL(n, \mathbf{C})$ is a subgroup of $U(n)$ and therefore bounded. $U(n)$ and $SL(n, \mathbf{C})$ are closed from which the compactness of $SU(n)$ follows.

The full unitary (orthogonal) group consists of two components and one such component is the special unitary (orthogonal) group.

Theorem 2.24   1. $U(n)$ and $U(1) \times SU(n)$ are homeomorphic.

2. $O(n)$ and $\{-1, 1\} \times SO(n)$ are homeomorphic.

In the unitary case the mapping is given by $(\det U, U_1) \mapsto U$ where U_1 is obtained from U by dividing the first column of U by $\det U$. The determinant $\det U$ is in $U(1)$ since $1 = \det E_n = \det(U^*U) = \det U^* \det U = |\det U|^2$. In the real case we observe that $\det R \in \{-1, 1\}$ and the rest follows by a similar argument as above.

The group $SL(2, \mathbf{R})$ is locally compact but no longer compact. It contains, for example, all matrices of the form $\begin{pmatrix} t & 0 \\ 0 & 1/t \end{pmatrix}$ with $t \in \mathbf{R}_0$. $SL(2, \mathbf{R})$ contains thus a group that is homeomorphic to $\mathbf{R}_0$. But from the Heine-Borel theorem it follows easily that this set is not compact since it is not bounded.

Let us now consider the case $n = 2$ in more detail. In this case we can calculate the form of the elements in $SU(2)$ and $SO(2)$.

Theorem 2.25   1. $SU(2)$ consists of all matrices of the form

$$U = \begin{pmatrix} \overline{b} & -\overline{a} \\ a & b \end{pmatrix}$$

with complex constants a and b that satisfy $\|a\|^2 + \|b\|^2 = 1$

2. $SO(2)$ consists of all matrices of the form

$$R = \begin{pmatrix} x & y \\ -y & x \end{pmatrix}$$

with real constants x, y and $x^2 + y^2 = 1$

This can be seen by a direct calculation. In the $SO(2)$ case the constants x and y are of course given by $x = \cos\phi$ and $y = \sin\phi$ for some angle ϕ.

From this theorem we find also that $SU(2)$ is topologically equivalent to the unit sphere in $\mathbf{R}^4$ and $SO(2)$ is topologically equivalent to the unit circle:

Theorem 2.26   1. Describe an element U in $SU(2)$ by the matrix

$$U = \begin{pmatrix} \overline{b} & -\overline{a} \\ a & b \end{pmatrix}$$

and set $a = x_1 + ix_2$ and $b = x_3 + ix_4$.
The mapping $f : SU(2) \rightarrow \mathbf{R}^4; U \mapsto (x_1, \quad x_2, \quad x_3, \quad x_4)$ is a homeomorphism of $SU(2)$ onto the unit sphere in $\mathbf{R}^4$.

2. The mapping $f : SO(2) \rightarrow U(1); R \mapsto e^{i\phi}$ is an topological isomorphism. Observe that $U(1)$ is identical to the unit circle in $\mathbf{C}$.

Next we will try to describe $SO(3)$ in different ways. When we introduced $SO(n)$ we viewed it as a subset of $\mathbf{R}^{n^2}$ that was defined by the equations $R'R = E_n$. $SO(3)$ is therefore the set of all solutions of the nine equations $r'_i r_j = \delta_{ij} (1 \leq i, j \leq 3)$ where r_i is the i-th column of R and r'_i is the i-th row. This is however not a very useful characterization. In the next theorem we will therefore characterize it with the help of three parameters, the so-called Euler angles:

Theorem 2.27 1. For every rotation $R \in SO(3)$ there are three angles φ_1, φ_2 and θ such that $R = R(\varphi_2)R(\theta)R(\varphi_1)$ where $R(\varphi)$ is a matrix of the form:

$$R(\varphi) = \begin{pmatrix} \cos\varphi & -\sin\varphi & 0 \\ \sin\varphi & \cos\varphi & 0 \\ 0 & 0 & 1 \end{pmatrix}$$

and $R(\theta)$ has the form:

$$R(\theta) = \begin{pmatrix} 1 & 0 & 0 \\ 0 & \cos\theta & -\sin\theta \\ 0 & \sin\theta & \cos\theta \end{pmatrix}$$

Furthermore: $0 \le \varphi_1, \varphi_2 \le 2\pi$ and $0 \le \theta < \pi$.

2. A rotation matrix R has the form:

$$R = \begin{pmatrix} \cos\varphi_1 \cos\varphi_2 - \cos\theta \sin\varphi_1 \sin\varphi_2 \\ \sin\varphi_1 \cos\varphi_2 + \cos\theta \cos\varphi_1 \sin\varphi_2 \\ \sin\varphi_2 \sin\theta \end{pmatrix}$$

$$\begin{pmatrix} -\cos\varphi_1 \sin\varphi_2 - \cos\theta \sin\varphi_1 \cos\varphi_2 & \sin\theta \sin\varphi_1 \\ \sin\varphi_1 \sin\varphi_2 + \cos\theta \cos\varphi_1 \cos\varphi_2 & -\sin\theta \cos\varphi_1 \\ \cos\varphi_2 \sin\theta & \cos\theta \end{pmatrix}$$

Assume that under the rotation R the axes x, y, z go to x', y', z'. By l we denote the intersection of the xy-plane with the $(x'y')$-plane. The angle between the x-axis and l is φ_1 and the angle between l and x' is φ_2, finally the angle between the z- and the z'-axis is θ. In the first step we rotate around the z-axis so that x goes to l. Then we rotate around l to move z to z'. Finally we rotate around z' so that x goes to x' and y to y'.

Theorem 2.28 1. A rotation $R \in SO(3)$ can be described by its axis and the rotation angle (The rotation axis is a space of vectors x with $Rx = x$).

2. Assume $1, e^{i\theta}, e^{-i\theta}$ are the eigenvalues of R. The rotation angle of R is then given by $\cos\theta = (\text{trace}(R) - 1)/2$.

3. For the rotation axis $v = (v_1, v_2, v_3)$ we find the ratio:

$$v_1 : v_2 : v_3 = R_{23} - R_{32} : R_{31} - R_{13} : R_{12} - R_{21}$$

To see this let λ_i and $v_i (i = 1, 2, 3)$ be the eigenvalues and eigenvectors of $R : Rv_i = \lambda_i v_i$. Since R is orthogonal we find that all eigenvalues have magnitude one: $|\lambda_i| = 1$. For the eigenvalues we find $\lambda_1 \lambda_2 \lambda_3 = \det R = 1$. Since R is a real matrix we find that the complex eigenvalues are conjugate complex. If there are complex eigenvalues (for example λ_2 and λ_3) then we find that the real eigenvalue λ_1 has value one. If there are only real eigenvalues then we find that $\lambda_i \in \{-1, 1\}$. We can thus find at least one eigenvalue equal to one. For this eigenvalue and the corresponding eigenvector we find $v = Rv$. v describes the rotation axis of R. Now let $\lambda_1 = 1, \lambda_2 = e^{i\theta}, \lambda_3 = e^{-i\theta}$ be the eigenvalues with eigenvectors v_1, v_2, v_3. These three vectors form an orthonormal set in $\mathbf{R}^3$ and we can go over to the new orthonormal system u_1, u_2, u_3 defined as

$$u_1 = v_1$$

$$u_2 = (v_1 + v_2)/\sqrt(2)$$

$$u_3 = i(v_1 - v_2)/\sqrt(2)$$

In this new coordinate system we find:

$$Ru_1 = u_1$$

$$Ru_2 = \cos\theta u_2 + \sin\theta u_3$$

$$Ru_3 = -\sin\theta u_2 + \cos\theta u_3.$$

The vector u_1 describes then the rotation axis and θ the rotation angle. The formula for the rotation angle follows from the fact that the trace of the matrix is equal to the sum of its eigenvalues.

The ratio of the components of the rotation axis can be found as follows: v is the rotation axis and therefore $Rv = v$. R is orthogonal and we get $R'R = E_n$, hence $v = R'v$ and therefore $(R - R')v = 0$ from which we get the ratio.

Using the mapping $R \mapsto (v_1\sin\theta, v_2\sin\theta, v_3\sin\theta, \cos\theta)$ where $v = (\, v_1 \quad v_2 \quad v_3 \,)$ is the rotation axis and θ the rotation angle of R we see that $SO(3)$ can be identified with the unit sphere in $\mathbf{R}^4$ where antipodal points are identified.

Finally we mention the following theorem about normal subgroups of $SO(3)$:

Theorem 2.29 $SO(3)$ is simple, i.e. it has only trivial normal subgroups.

2.3.3 Invariant Integration

In section 2.1.3 we sketched very briefly one way to construct a generalized Riemann integral in $\mathbf{R}^n$. In the following we need however an integral on arbitrary locally compact topological groups. In this section we will introduce such an integral. In contrast to the previous generalization this time we will not construct the integral. Instead we will list some of its properties and then we will formulate a theorem that ensures the existence of such an integral.

Definition 2.46 Let X be a topological space. Then we define:

1. The *support* of a real valued function $f : X \to \mathbf{R}$ is the closure of the set of all $x \in X$ where f does not vanish. We write:

$$\text{support}(f) = \{x \in X | f(x) \neq 0\}\,.$$

2. $C_0(X)$ is the space of all continuous, real functions on X with compact support.

3. $C_0^+(X)$ is the subset of the non-negative functions in $C_0(X)$.

4. A real-valued function on $C_0^+(X)$ is called a *functional*. A functional is thus a mapping that maps a function to a real number.

Now we assume that the space X is a locally compact group G. We consider the transformation of functions f on G under the group operation:

Definition 2.47   1. For a function $f : G \to \mathbf{R}$ and a fixed element $a \in G$ we define the *left-translate* $^a f$ and the *right-translate* f^a as:

$$^a f(x) = f(ax) \qquad f^a(x) = f(xa)$$

2. A functional I is called *left-invariant*, *(right-invariant)* if $I(^a f) = I(f)$ $(I(f^a) = I(f))$ for all $a \in G$ and all $f \in C_0^+(G)$.

3. A functional I is called *non-negative* if $I(f) \geq 0$ for all $f \in C_0^+$.

4. A functional I is called *positive-homogeneous* if $I(\lambda f) = \lambda I(f)$ for all $f \in C_0^+(G)$ and all $\lambda \geq 0$.

For functionals we have the following existence and uniqueness properties:

Theorem 2.30   1. Let G be a locally compact topological group. Then there exists a non-trivial (i.e. not identically zero), non-negative, left-invariant, positive homogeneous, additive functional on $C_0^+(G)$.

2. This functional is unique up to a multiplicative constant, i.e. if I and J are functionals with the above mentioned properties, then there is a constant c such that $I(f) = cJ(f)$ for all $f \in C_0^+(G)$.

3. There is also a right-invariant functional with the same properties.

Definition 2.48 The integral constructed in the previous theorem is called the *Haar Integral* of the group.

These invariant functionals on $C_0^+(G)$ can be extended to real functions in $C_0(G)$ and complex functions:

Theorem 2.31   1. The left and right-invariant functionals can be extended to functionals on $C_0(G)$ with the following construction: For an $f \in C_0(G)$ we define the positive part f^+ of f as:

$$f^+(g) = \begin{cases} f(g) & \text{if } f(g) \geq 0 \\ 0 & \text{else} \end{cases}$$

In a similar way one can define f^-, the negative part of f. For a $f \in C_0(G)$ we define: $I(f) = I(f^+) - I(f^-)$.

2. For complex valued functions $f = f_r + i f_i$ we define $I(f) = I(f_r) + i I(f_i)$.

2.3.4   Examples of Haar Integrals

We will now derive Haar integrals for some important groups. First we give a general theorem that describes the Haar integral with the help of the Jacobian.

Theorem 2.32 Assume that $X \subset \mathbf{R}^n$ is an open subset of $\mathbf{R}^n$ and that it satisfies the following conditions:

1. Assume that there is a product $\circ$ on X under which X becomes a topological group. We denote the function describing the product by
   $$f : \mathbf{R}^{2n} \to \mathbf{R}^n; (x, y) \mapsto x \circ y$$

2. Assume p_i is the projection to the coordinate axis: $p_i(x_1, ..., x_n) = x_i$. Then we assume that $f_i = p_i \circ f$ is continuously differentiable.

3. The Jacobians $J(l_a)$ and $J(r_a)$ of the left- and right-translations: $l_a(x) = f(a, x)$ and $r_a(x) = f(x, a)$ are constant.

Under these conditions the left and right Haar integrals are given by the functionals:

$$I_l(f) = \int_X f(x) \, |J(l_x)|^{-1} \, dx$$

$$I_r(f) = \int_X f(x) \, |J(r_x)|^{-1} \, dx$$

This follows directly from the chain rule.

The following examples of Haar integrals can be computed with this theorem:

Examples 2.9   1. Set $X = \mathbf{R}_0$ and define $f(x, y) = xy$. The Jacobian is given by $J(a) = a$. The left and right Jacobian are identical. Left and the right Haar integral are therefore also identical and we have

$$\int_{\mathbf{R}_0} f(g) \, dg = \int_{\mathbf{R}_0} \frac{f(x)}{|x|} \, dx$$

2. Now use the multiplicative complex group instead of the reals We get $f(\xi, \eta) = f(a + ib, x + iy) = \xi\eta = ax - by + i(ay + bx)$. The Jacobian is given by the matrix $\begin{pmatrix} a & -b \\ b & a \end{pmatrix}$ and we find for the Haar integral:

$$\int_{\mathbf{C}_0} f(g) \, dg = \int_{-\infty}^{\infty} \int_{-\infty}^{\infty} \frac{f(x + iy)}{x^2 + y^2} \, dx \, dy$$

3. Define X to be the set of matrices $\begin{pmatrix} x & y \\ 0 & 1 \end{pmatrix}$ with a positive real number x and arbitrary real y. This set is a locally compact group under the usual matrix multiplication. For a fixed element $\begin{pmatrix} a & b \\ 0 & 1 \end{pmatrix}$ we find

$$\begin{pmatrix} x & y \\ 0 & 1 \end{pmatrix} \begin{pmatrix} a & b \\ 0 & 1 \end{pmatrix} = \begin{pmatrix} ax & bx + y \\ 0 & 1 \end{pmatrix}$$

and

$$\begin{pmatrix} a & b \\ 0 & 1 \end{pmatrix} \begin{pmatrix} x & y \\ 0 & 1 \end{pmatrix} = \begin{pmatrix} ax & ay + b \\ 0 & 1 \end{pmatrix}$$

and the Jacobians are

$$\begin{pmatrix} a & b \\ 0 & 1 \end{pmatrix} \quad \text{and} \quad \begin{pmatrix} a & 0 \\ 0 & a \end{pmatrix}.$$

We get therefore $|J(l_a)| = a^2$ and $|J(r_a)| = a$. The left- and the right Haar integral are thus different and given by

$$\int_X \frac{f(x,y)}{x^2} \, dx dy \quad \text{and} \quad \int_X \frac{f(x,y)}{x} \, dx$$

respectively.

4. In the Euler angles the invariant integral on $SO(3)$ is given by

$$\int_{SO(3)} f(g) \, dg = \frac{1}{8\pi^2} \int_0^{2\pi} \int_0^{\pi} \int_0^{2\pi} f(\phi, \theta, \psi) \sin \theta d\phi d\theta d\psi$$

A derivation of the last integral can be found in [15].

2.3.5　Exercises

Exercise 2.22 Show that $GL(n, \mathbf{R})$ and $GL(n, \mathbf{C})$ are open sets in $\mathbf{R}^{n^2}$ and $\mathbf{C}^{n^2}$ respectively.

Exercise 2.23 A neighborhood $U(e)$ of the identity is called a *symmetrical neighborhood* if $U(e)^{-1} = U(e)$. Show that every neighborhood of the identity contains a symmetrical neighborhood. (Hint: $g \mapsto g^{-1}$ is a homeomorphism.)

Exercise 2.24 Show that every neighborhood $U(e)$ of the identity contains a neighborhood $V(e)$ such that $V(e)V(e) \subset U(e)$. (Hint: $f(gh) = gh$ is continuous.)

Exercise 2.25 Denote by $S(a, b)$ the matrix $\begin{pmatrix} \bar{b} & -\bar{a} \\ a & b \end{pmatrix}$ and by $R(a, b)$ the matrix:

$$\begin{pmatrix} (b^2 - a^2 + \bar{b}^2 - \bar{a}^2)/2 & i(b^2 + a^2 - \bar{b}^2 - \bar{a}^2)/2 & ab + \overline{ab} \\ i(-\bar{b}^2 + \bar{a}^2 + b^2 - a^2)/2 & (b^2 + a^2 + \bar{b}^2 + \bar{a}^2)/2 & i(ab - \overline{ab}) \\ -ab - \overline{ab} & i(\overline{ab} - ab) & b\bar{b} - a\bar{a} \end{pmatrix}$$

Show the the mapping $f : SU(2) \to SO(3)$ defined as $f(S(a, b)) = R(a, b)$ is an open continuous homomorphism of $SU(2)$ onto $SO(3)$. Show that the kernel is given by the matrices E_2 and $-E_2$.

　　　Hint: Investigate first

$$f\left(\begin{pmatrix} e^{-(i\varphi)/2} & 0 \\ 0 & e^{i\varphi/2} \end{pmatrix} \right)$$

and

$$f\left(\begin{pmatrix} \cos(\theta/2) & i\sin(\theta/2) \\ i\sin(\theta/2) & \cos(\theta/2) \end{pmatrix} \right)$$

Exercise 2.26 Assume V is a finite-dimensional, real vector space and $\langle \, , \, \rangle : V \times V \to$ **R** is a bilinear map. Define

$$G = \{A | A : V \to V, \langle v, w \rangle = \langle Av, Aw \rangle \text{ for all } v, w \in V\} .$$

- Show that G is a group.
- Assume $V = \mathbf{R}^n, H = \mathbf{R}$ and $\langle \, , \, \rangle$ is the standard scalar product. What is G in this case?

Exercise 2.27 The 2-D rotations transform circles into circles, the n-dimensional rotations map n-dimensional spheres into themselves. The circles and spheres are thus invariant under the elements of $SO(2)$ and $SO(n)$ respectively.

Now consider the vector space $V = \mathbf{R}^2, H = \mathbf{R}$ and the mapping

$$\langle x, y \rangle = x_1 y_1 - x_2 y_2.$$

Assume further that L is a function $L : V \to V$ that leaves this mapping invariant: $\langle v, w \rangle = \langle Lv, Lw \rangle$. What can be said of the invariant subspaces of L?

(Remark: Transformations that leave functions of the form $\langle x, y \rangle = \sum_{k=1}^{n-1} x_k y_k - x_n y_n$ invariant are called *Lorentz groups*.)

Chapter 3

Representations of Groups

In this chapter we will introduce the concept of a representation of a group. The methods introduced here are of fundamental importance in the study of symmetries and they have been applied in such different fields as quantum mechanics and the theory of special functions. Apart from these "applications" the theory is also useful for the study of abstract groups since these abstract groups are linked via representations to well-known matrix groups.

In the next section we will first introduce the algebraic concepts involved and then we will specialize the concepts to the case where the groups also have a topological structure.

3.1 Algebraic Representations

Definition 3.1 Let G be any group, X a complex vector space and denote by $GL(X)$ the group of invertible, linear mappings that carry X into itself.

1. Then we define a *representation* as a homomorphism
   $T : G \to GL(X); g \mapsto T(g)$.

2. X is called the *representation space.*

3. $T(g)$ is called a *representation operator.*

We call a representation in the sense of definition 3.1 an *algebraic* representation in contrast to the *continuous* representations to be introduced in the next section 3.12 (see also [37] page 151). In the definition of continuous representations we will mainly require that the operators T are continuous, a property which is of course coupled to the topological properties of the group and the vector space.

The vector space is finite-dimensional in almost all applications of interest to us. In this case we define:

Definition 3.2   1. If the representation space X is a finite-dimensional vector-space then we call the representation *finite* or *finite-dimensional.* Otherwise it is called *infinite* or *infinite-dimensional.*

2. The dimension of X is called the *dimension of the representation.*

3. Assume T is an n-dimensional representation and $e_1, \ldots, e_n$ is an arbitrary but fixed basis of X. Relative to this basis $T(g)$ can be described by a matrix

$$t(g) = \begin{pmatrix} t_{11}(g) & \cdots & t_{1n}(g) \\ \vdots & & \vdots \\ t_{n1}(g) & & t_{nn}(g) \end{pmatrix}$$

$t(g)$ is called the *matrix of the representation* and the functions $t_{ij}(g)$ are called its *matrix elements*.

If we do not say otherwise, then the representations are assumed to be finite-dimensional. From the definition we find the following properties of a representation:

Theorem 3.1   1. $T(e) = id$, where e is the identity in G and id the identity mapping in $L(X)$.

2. $T(g^{-1}) = T(g)^{-1}$

The first property is a general property of a homomorphism. To see the second equation compute: $id = T(e) = T(gg^{-1}) = T(g)T(g^{-1})$.

For the corresponding matrices we get:

Theorem 3.2   1. If t is a matrix representation of G then we have the matrix equation:

$$t(g_1g_2) = t(g_1)t(g_2).$$

2. For the matrix elements this becomes:

$$t_{ij}(g_1g_2) = \sum_k t_{ik}(g_1)t_{kj}(g_2)$$

3. The matrix $t(e)$ is the identity matrix.

By an easy calculation it can be shown that the following functions are examples of one-dimensional representations:

Examples 3.1   1. Let $G = \mathbf{R}$ be the additive real group and $k \in \mathbf{C}$. The function: $T_k(x) = e^{kx}$ is a one-dimensional representation. We write e^{kx} instead of $\left(e^{kx}\right)$ but we have to remember that $T_k(x)$ is a function $T_k(x) : \mathbf{C} \to \mathbf{C}$ with $T_k(x)(\zeta) = e^{kx}\zeta$.

2. Let $G = \mathbf{C}$ be the additive complex group and let $k_1, k_2 \in \mathbf{C}$ be two complex constants. Then $z = x + iy \mapsto e^{k_1x}e^{k_2y}$ is a one-dimensional representation. As in the previous case we have: $T_{k_1k_2}(x + iy)\zeta = e^{k_1x}e^{k_2y}\zeta$.

3. Let $G = \mathbf{R}_0$ be the non-zero multiplicative real group, $k \in \mathbf{C}$ be a complex constant and $\epsilon \in \{0,1\}$. Then $x \mapsto e^{k(\ln|x|)}(\text{sign } x)^\epsilon$ is a one-dimensional representation of $\mathbf{R}_0$.

4. Let $G = SO(2)$ be the circle group, ϕ be the rotation angle and $n \in \mathbf{Z}$. Then $\phi \mapsto e^{in\phi}$ is a one-dimensional representation of $SO(2)$.

A very important property of representations is connected to subspaces of X which are left invariant under all transformations $T(g)$:

Definition 3.3   1. A subspace $Y \subset X$ is called an *invariant subspace of* T if Y is left invariant under all transformations $T(g), g \in G$, i.e.

$$T(g)y \in Y \text{ for all } g \in G \text{ and all } y \in Y$$

2. A representation is called *irreducible* if $\emptyset$ and X are the only invariant subspaces of X.

3. A representation is *reducible* if it is not irreducible.

By induction it is easy to prove that if T is a representation on a finite-dimensional space X then there is a subspace $Y \subset X$ on which T is irreducible.

There are of course infinitely many representations for each group but many of them are essentially equal to each other. In the next definition we specify what we mean by essentially equal representations:

Definition 3.4 Assume S and T are two representations of a group G in the spaces X and Y. S and T are called *equivalent* $(T \sim S)$ if there is an isomorphism $A : X \to Y$ such that $AT(g) = S(g)A$ for all $g \in G$.

The operator A in the previous definition is only one example of a function that connects two representations. In the general case we define:

Definition 3.5 Let S and T be representations in the spaces X and Y respectively. Assume further that $A : X \to Y$ is a linear operator. We say that A is an *intertwining operator for* T *and* S if $AT(g) = S(g)A$ for all $g \in G$.

An intertwining operator is obviously a generalization of the equivalence concept since S and T are equivalent representations in the case where A is an isomorphism and where A is an intertwining operator for S and T.

For finite-dimensional representations (and especially one-dimensional representations) we get the following characterization of equivalent representations:

Theorem 3.3   1. Two finite-dimensional representations are equivalent if there are two bases in X and Y in which their matrices coincide (This means especially that X and Y have the same dimension).

2. Two finite-dimensional matrix representations t and s are equivalent if there is a matrix A such that $At(g)A^{-1} = s(g)$ for all $g \in G$.

3. Equivalent one-dimensional representations are defined by the same function $t(g)$.

Irreducible representations can be characterized by intertwining operators. This result is known as Schur's Lemma:

Theorem 3.4   1. Assume S and T are irreducible representations of the group G in the spaces X and Y. Assume further that $A : X \to Y$ is an intertwining operator for T and S (i.e. $AT(g) = S(g)A$ for all $g \in G$). Then A is either an isomorphism or $A = 0$.

2. Assume T is a finite-dimensional, irreducible representation of G in X. Then every linear operator $B : X \to X$ that satisfies: $BT(g) = T(g)B$ for all $g \in G$ has the form $B = \lambda id$ where λ is a complex constant.

3. An irreducible, finite-dimensional representation of a commutative group is one-dimensional.

Proof: The image $L = AX$ of X under A is a subspace of Y. L is invariant under $S(g)$ because $S(g)Ax = A(T(g))x = Ax' \in L$. S is irreducible and L is therefore the null-space or the whole space Y. In the first case $A = 0$, in the second case we find that A is surjective and we show that A is also injective. We note that the kernel is an invariant subspace and therefore it is either equal to the null-space or to the whole space. The second case is impossible since A would otherwise be the null-space. This proves the first part of the theorem.

To show the second part we note that B is a linear operator and therefore it has at least one eigenvalue λ. We define $A = B - \lambda id$. A is intertwining but it is not an isomorphism and therefore A is the null-operator by the previous part of the theorem.

If T is an irreducible, finite-dimensional representation of the commutative group G then $T(g_0)$ satisfies

$$T(g)T(g_0) = T(gg_0) = T(g_0g) = T(g_0)T(g)$$

for all $g \in G$. Therefore we find that $T(g_0) = \lambda_0 id$ by the previous part. Every subspace of X is therefore invariant under all $T(g)$ and the dimension of the space must therefore be one.

We will now describe how we can decompose a given representation into a number of simpler components. For this purpose we introduce the direct sum of representations:

Definition 3.6 Assume G is a group and $X_1, \ldots, X_n$ are vector spaces. Assume further that T^i are representations of G in X_i. Then we define the *direct sum* T of the representations T^i as follows: As representation space we take the direct sum of the vector spaces and as mapping we define $T(g)(x) = T(g)(x_1 + \ldots + x_n) = T^1(g_1)x_1 + \ldots + T^n(g_n)x_n$.

Now assume that all the representations are finite-dimensional. Then we select a basis in each vector space and get a basis in the sum by the union of all the basis elements. If we group them together in a correct order then we find that the matrix representation of the sum is given by

$$t(g) = \begin{pmatrix} t^1(g) & & & & & \\ & t^2(g) & & & & \\ & & t^3(g) & & & \\ & & & \cdot & & \\ & & & & \cdot & \\ & & & & & t^n(g) \end{pmatrix}$$

We introduced the direct sum of representations as a tool to build a new representation from a number of given representations. In the following definition we invert this process by introducing a special class of representations that can be decomposed into a number of simpler components.

Definition 3.7 A representation is called *completely reducible* if it is the direct sum of a finite number of irreducible representations .

The matrices of a finite-dimensional, completely reducible representation can thus be simultaneously diagonalized so that the diagonal matrices define irreducible matrix representations.

A reducible representation is not necessarily completely reducible as the following example shows:

Examples 3.2 Take as group the real numbers **R** and as representation the mapping $x \mapsto T(x) = \begin{pmatrix} 1 & x \\ 0 & 1 \end{pmatrix}$, i.e. each number is mapped into a translation. Then we find for an element $\begin{pmatrix} \xi_1 \\ \xi_2 \end{pmatrix}$ of an one-dimensional invariant subspace M the equations:

$$T(x) \begin{pmatrix} \xi_1 \\ \xi_2 \end{pmatrix} = \begin{pmatrix} \xi_1 + x\xi_2 \\ \xi_2 \end{pmatrix} = \lambda(x) \begin{pmatrix} \xi_1 \\ \xi_2 \end{pmatrix} = \begin{pmatrix} \lambda(x)\xi_1 \\ \lambda(x)\xi_2 \end{pmatrix}$$

and from this we find that the only invariant subspace M is the space given by the vectors $\begin{pmatrix} \xi_1 \\ 0 \end{pmatrix}$. The representation is thus reducible but not completely reducible.

Up to now we used only the basic properties of groups and vector spaces. In the next definition we will now assume that the representation space X is also equipped with a scalar product $\langle \, , \, \rangle$ and we will study representations which are compatible with the scalar product, i.e. representations that leave the geometry in X invariant.

Definition 3.8 Assume T is a representation of the group G in X. A representation T^* is called the *adjoint representation* if we have for all $x, y \in X$ and all $g \in G$

$$\langle T(g)x, T^*(g)y \rangle = \langle x, y \rangle .$$

For the matrices of adjoint representations we find the following characterization:

Theorem 3.5 Assume T is a finite-dimensional representation with matrix representation $t(g) = (t_{ij}(g))$. Then the adjoint representation $S(g)$ has the matrix representation $s(g) = (s_{ij}(g))$ with $t_{ij}(g^{-1}) = \overline{s_{ji}(g)}$.

Of special importance are the representations that leave the metrical properties of the vector spaces invariant. These representations are the selfadjoint, or unitary representations:

Definition 3.9 1. Assume X is a pre-Hilbert space with a scalar product $\langle \, , \, \rangle$. Then we say that a representation T is *unitary* if

$$\langle T(g)x, T(g)y \rangle = \langle x, y \rangle$$

for all $g \in G$ and all $x, y \in X$.

2. An $n \times n$ matrix t is *unitary* if $t^*t = E_n$ where t^* is the transposed, conjugate complex of t.

By a simple calculation we find:

Theorem 3.6   1. The matrices $t(g)$ of a unitary representation T in an orthonormal basis are unitary.

2. If $t(g)$ is the matrix of the unitary representation T then we have:

$$t^*(g) = t(g)^{-1} = t(g^{-1})$$

$$t_{ij}(g^{-1}) = \overline{t_{ji}(g)}$$

If a vector space is not a pre-Hilbert space or if the original scalar product is not compatible with a given representation then it is sometimes possible to introduce a new scalar product under which the representation is unitary. In this case we define:

Definition 3.10 A representation T in a space X is called *equivalent to a unitary representation* if there is a scalar product on X under which T becomes a unitary representation.

For these representations we have:

Theorem 3.7 If a finite-dimensional representation is equivalent to a unitary representation then it is completely reducible.

Proof: If M is an invariant subspace then $M^\perp$ is also invariant. Furthermore we know that any finite representation has an invariant subspace on which T is irreducible. Assume M_1 is such a subspace. If $M_1 = X$ we are through, otherwise we get $M_1^\perp \neq \{0\}$ and $M_1^\perp$ is also invariant under T. Now do the same construction on M_1 instead of M and get a subspace M_2 on which T is irreducible. If M_1 and M_2 span X then we are through, otherwise we do the same for the remaining space.

As a last construction in this section we introduce the character of a representation. It will be used later on (see section 5.2) as a tool to characterize representations. Here we derive only the basic properties of the character.

Definition 3.11   1. If $t = (t_{ij})$ is an $n \times n$−matrix then we define the *trace* of t as: $tr(t) = \sum_{i=1}^{n} t_{ii}$. The trace is thus the sum of all diagonal elements.

2. Assume that T is a finite-dimensional representation and t its matrix representation. We define the *character of the representation* as the function $\chi_T : G \to \mathbf{C}; g \mapsto \chi_T(g) = tr(t(g))$.

Note the matrix representation t of T depends on the selected basis of X but since a change of bases with a transformation matrix A results in the equivalent matrix representation $A^{-1}tA$ and since $tr(A^{-1}tA) = tr(t)$ we find the χ_T is independent of the basis chosen.

From the properties of the trace we find with some simple computations:

Theorem 3.8 For finite-dimensional representations we have the following properties:

1. If T and S are equivalent representations then $\chi_T = \chi_S$.

2. Characters are constant on conjugacy classes in G.

3. If T is unitary then $\chi_T(g^{-1}) = \overline{\chi_T(g)}$.

Proof: The first part was already mentioned in the definition of a character. For the second we note that g_1 and g_2 are conjugate if there is a $g \in G$ such that $g_2 = gg_1g^{-1}$ and we get: $\chi_T(g_2) = tr(t(g_2)) = tr(t(gg_1g^{-1})) = tr(t(g)t(g_1)t(g^{-1})) = tr(t(g)t(g_1)t(g)^{-1}) = tr(t(g_1)) = \chi_T(g_1)$.

If T is unitary then we have $t(g^{-1}) = t^*(g)$ and since we have $tr(t^*) = \overline{tr(t)}$ we see at once the last part of the theorem.

3.2 Continuous Representations

In the previous section we introduced representations of groups in vector spaces and pre-Hilbert spaces. In this section we will concentrate on the case where G is a topological group and where X is a topological vector space or even a Hilbert space.

Definition 3.12   1. A *topological vector-space* is a vector space which is also a Haussdorf topological space and which satisfies the following conditions:

   a) Addition is continuous, i.e. the mapping $(x, y) \mapsto x + y$ is a continuous function on $X \times X \to X$.

   b) Scalar multiplication is continuous, i.e. $(c, x) \mapsto cx$ is a continuous function on $\mathbf{C} \times X \to X$.

2. Assume G is a topological group and X is a topological vector space. A mapping $T : G \to GL(X)$ is a *continuous representation* if T is an algebraic representation and if $(x, g) \mapsto T(g)x$ is a continuous mapping of $X \times G \to X$.

3. Assume S and T are two continuous representations in two spaces X and Y then we call S and T *equivalent* if there is a homeomorphism $A : X \to Y$ which is linear and one-to-one and which satisfies:

$$AS(g) = T(g)A$$

for all $g \in G$.

4. A continuous representation T is called *irreducible* if $\{0\}$ and X are the only closed, invariant subspaces of T.

We will in the following only speak of representations when we mean continuous representations. The definition of representations and algebraic representations are identical up to the following, more or less technically motivated, differences:

1. The mapping T is continuous in g and x

2. The intertwining operator A in the equivalence definition must be a homeomorphism and

3. in the definition of irreducible representations we added the condition that the invariant subspaces must be closed.

From the definition we see direct:

Theorem 3.9 The matrix elements $t_{ij}(g)$ and the character $\chi_T(g)$ are complex-valued, continuous functions on G.

We can now show that representations of compact groups on Hilbert spaces have an especially simple structure, they are completely reducible as will be shown in the following two theorems:

Theorem 3.10 A representation T of a compact group in a Hilbert space H is equivalent to a unitary representation.

We note that there is a Haar integral $\int_G f(g)\ dg$ on G since G is compact. We have to show that H can be equipped with a scalar product under which T is unitary. It is easy to see that such a scalar product is given by

$$\langle x, y \rangle = \int_G \langle T(g)x, T(g)y \rangle_1\ dg$$

where $\langle x, y \rangle_1$ is the original scalar product in H.

For unitary representations one can show that the orthogonal complement of an invariant subspace is also invariant and we get:

Theorem 3.11 All finite-dimensional representations of a compact group are completely reducible.

The next theorem, stated without proof, shows that it is sufficient to consider only finite-dimensional, irreducible, unitary representations of a compact group (see [37]):

Theorem 3.12 All irreducible, unitary representations of a compact group are finite-dimensional.

Up to now we derived a number of properties of representations and we listed some examples of algebraic representations. In the following theorem we will now construct all irreducible representations of the group of additive, real numbers $\mathbf{R}$. This group is commutative and we know therefore from Schur's lemma (see 3.4) that these representations are all one-dimensional. It is therefore sufficient to construct all one-dimensional representations of $\mathbf{R}$. This is done in the next theorem:

Theorem 3.13 All one-dimensional representations $x \mapsto f(x)$ of the group $\mathbf{R}$ are given by $f(x) = e^{kx}$ with a complex constant $k \in \mathbf{C}$.

Note again that e^{kx} stands for the 1×1 matrix $\left(e^{kx} \right)$.

Proof: Assume that f is a one-dimensional representation of $\mathbf{R}$. From the properties of representations we find the following equations for f:

$$f(0) = 1, \tag{3.1}$$

$$f(x + y) = f(x)f(y) \text{ for all } x, y \in \mathbf{R}. \tag{3.2}$$

and

$$f(-x) = f^{-1}(x) \text{ for all } x \in \mathbf{R} \tag{3.3}$$

From the last equation 3.3 we get:

$$f(x) \neq 0 \text{ for all } x \in \mathbf{R}. \tag{3.4}$$

Now take a point $x_0 \in \mathbf{R}$ and a function $g(x)$ with the following properties:

- g is infinitely differentiable,

- g is zero outside a neighborhood of x_0 and

- there is a constant $c \neq 0$ such that:

$$\int_{-\infty}^{\infty} f(x)g(x) \, dx = c \neq 0. \tag{3.5}$$

The existence of such a function follows from equation 3.4. From equation (3.5) we find

$$\int_{-\infty}^{\infty} f(x+y)g(y) \, dy = f(x) \int_{-\infty}^{\infty} f(y)g(y) \, dy = f(x) \cdot c$$

and

$$f(x) = \frac{1}{c} \int_{-\infty}^{\infty} f(x+y)g(y) \, dy = \frac{1}{c} \int_{-\infty}^{\infty} f(y)g(y-x) \, dy$$

From the properties of g it follows that the last integral is infinitely differentiable. From equation (3.2) we conclude that f is a solution of the differential equation:

$$f'(x) = kf(x).$$

We conclude that $f(x) = e^{kx}$ for some complex constant k.

The previous theorem shows that the exponential function is characterized by the transformation property $e^{x+y} = e^x e^y$. Another way to express the same fact is to say that the exponential function is characterized by its property of defining irreducible representations of $\mathbf{R}$. We saw also how irreducible representations define a link between groups and functions that behave nicely under the group operations. We found also in the case $G = \mathbf{R}$ that these functions are characterized by this transformation property. One application of the theory of group representations is thus the investigation of this type of problem in the general group theoretical setting, i.e. given a group G what are the special functions connected to G via group representations.

For other commutative groups we could derive similar results, all involving the exponential function in some way. In the following examples we give a list of irreducible representations of different commutative groups (see [37]):

Examples 3.3 1. All unitary, one-dimensional representations of the group $\mathbf{R}$ are given by $x \mapsto f(x) = e^{itx}$ with a real constant $t \in \mathbf{R}$.

2. All one-dimensional representations $x \mapsto f(x)$ of the group $\mathbf{R}^n$ are of the form $f(x_1, \ldots, x_n) = e^{k_1 x_1 + \ldots + k_n x_n}$ with complex constants $k_i \in \mathbf{C}$.

3. All unitary one-dimensional representations $x \mapsto f(x)$ of the group $\mathbf{R}^n$ are given by $f(x_1, \ldots, x_n) = e^{i(t_1 x_1 + \ldots + t_n x_n)}$ with real constants $t_i \in \mathbf{R}$.

4. All one-dimensional representations of the group $\mathbf{C}^n$ are given by

$$z \mapsto f(z) = f(z_1, \ldots, z_n) = e^{p_1 z_1 + q_1 \overline{z_1} + \ldots + p_n z_n + q_n \overline{z_n}}$$

with complex constants $p_i, q_i \in \mathbf{C}$.

5. All unitary one-dimensional representations $z \mapsto f(z)$ of the group $\mathbf{C}^n$ are given by $f(z_1, \ldots, z_n) = e^{p_1 z_1 - \overline{p_1 z_1} + \ldots + p_n z_n - \overline{p_n z_n}}$ with complex constants $p_i \in \mathbf{C}$.

6. All one-dimensional representations of the 2-D rotation-group $SO(2)$ are given by $\alpha \mapsto f(\alpha)$ with $f(\alpha) = e^{im\alpha}$ where $m \in \mathbf{Z}$ is an integer and $\alpha \in [0, 2\pi]$ is the rotation angle. All representations are unitary.

7. All one-dimensional representations $x \mapsto f(x)$ of the group $\mathbf{R}_0^+$ are given by $f(x) = x^k = e^{k \ln(x)}$ with a complex constant $k \in \mathbf{C}$.

8. All unitary one-dimensional representations $x \mapsto f(x)$ of the group $\mathbf{R}_0^+$ are given by $f(x) = x^k = e^{it \ln(x)}$ with a real constant $t \in \mathbf{R}$.

9. All one-dimensional representations $x \mapsto f(x)$ of the group $\mathbf{R}_0$ are given by $f(x) = (\text{sign} x)^\epsilon |x|^k$ with a complex constant $k \in \mathbf{C}$.

10. All one-dimensional representations of $\mathbf{C}_0^1$ are given by

$$f(z) = |z|^k \, e^{im \arg z}$$

with a complex constant $k \in \mathbf{C}$ and an integer $m \in \mathbf{Z}$.

11. All unitary one-dimensional representations of $\mathbf{C}_0^1$ are given by

$$f(z) = |z|^{it} \, e^{im \arg z}$$

with a real constant $t \in \mathbf{R}$ and an integer $m \in \mathbf{Z}$.

12. All one-dimensional representations of $\mathbf{C}_0^n$ are given by

$$f(z) = |z_1|^{k_1} \ldots |z_n|^{k_n} \, e^{i(m_1 \arg z_1 + \ldots m_n \arg z_n)}$$

with complex constants $k_i \in \mathbf{C}$ and integers $m_i \in \mathbf{Z}$.

13. All unitary one-dimensional representations of $\mathbf{C}_0^n$ are given by

$$f(z) = |z_1|^{it_1} \ldots |z_n|^{it_n} \, e^{i(m_1 \arg z_1 + \ldots m_n \arg z_n)}$$

with real constants $t_i \in \mathbf{R}$ and integers $m_i \in \mathbf{Z}$.

A natural way to construct representations is the following: Consider a group G with Haar integral $\int_G f(g) \, dg$. Then we define $L^2(G)$ as usual as the space of all square-integrable functions on G. This space is of course a vector space and seems to be natural to use it as the representation space for representations of G. A simple calculation shows that the mapping $g \mapsto T(g)$ with $T(g)f(h) = f(hg)$ defines indeed a representation of G in $L^2(G)$.

Finally we note that the representation space X together with the group $T(G) = \{T(g)|g \in G\}$ form a transformation group $(X, T(G))$. The orbit of an element $x \in X$ is given by the set $\{T(g)x|g \in G\}$. Sometimes we will call two elements x and $y \in G$ $T(G)$ equivalent if they are members of the same orbit, i.e. if there is a group element $g \in G$ such that $y = T(g)x$. In this case we write also $x \overset{T(G)}{\equiv} y$ or simply $x \equiv y$. $T(G)$ equivalence is an equivalence relation and we find therefore that the space X is the disjoint union of the orbits.

3.3 Exercises

Exercise 3.1 Prove the algebraic representation properties for some of the representations mentioned in the examples 3.1.

Exercise 3.2 Prove theorem 3.5.

Exercise 3.3 Prove some parts of theorem 3.3.

Exercise 3.4 Assume $k(X)$ is a kernel function that depends only on the magnitude of X, i.e. there is a function $h : \mathbf{R}_+ \to \mathbf{R}$ such that $k(X) = h(\|X\|)$.

Define the operator $K : L^2(\mathbf{R}^n) \to L^2(\mathbf{R}^n)$ by the convolution:

$$Kf(Y) = \int_{\mathbf{R}^n} k(X - Y)f(X)\, dX$$

show that K is an intertwining operator for the rotation group.

The operator given by $h(r) = Ce^{\frac{r^2}{\sigma^2}}$ defines Gaussian blur. In another example h might describe the optical properties of a circular lens, in this case f would describe the intensity distribution of the original object and Kf would be the intensity distribution of the recorded image.

Chapter 4

Representations of some Matrix Groups

In this chapter we will construct representations for a number of important groups. In the first section we will treat SO(3) in some detail. We do this since this group is of great practical importance and since it is the simplest compact, non-commutative group. By constructing the representations of SO(3) we will also demonstrate the usage of some of the fundamental tools of Lie group theory. We conclude the section on SO(3) by collecting a number of formulas that are important in 3-D image processing. Further practical information about the surface harmonics can be found in books on special functions like [11].

In the next two sections we will then consider the group of motions in the euclidian plane and the special linear group of 2×2 matrices. In the case of the group $M(2)$ we construct some irreducible representations of this group using the same techniques as in the SO(3) case.

A complete theory of representations for these groups is considerably more difficult then the representation theory for SO(3) and lies beyond the scope of these lecture notes. We will therefore only give a collection of basic facts about these groups. The interested reader can find detailed investigations of these groups in the literature. We mention here that a detailed treatment of the representation theory of the euclidian motion group $M(2)$ can be found in [45] (chapter IV, pages 149-198) and the representations of $SL(2, \mathbf{C})$ are investigated in [14] (chapter IV, pages 202-272).

4.1   The Representations of $SO(3)$

From the previous sections it should be clear that it is important to know a complete set of representations of a group. In this section we will now construct such a set for the group of rotations in 3-D. This is considerably more difficult than in the 2-D case since $SO(3)$ is not commutative. We can therefore expect that we will have to consider higher dimensional representations. Since $SO(3)$ is compact it is however only necessary to find all irreducible, unitary representations (see 3.11 and 3.10). We will now derive a complete set of such representations.The representation space is $L^2(S^2)$ the space of square integrable functions on the unit sphere.

This will lead us to a system of functions Y_l^m that is connected to the three-dimensional representations as the exponential functions are connected to the two-dimensional rotation group $SO(2)$.

Recall that the exponential functions are characterized by the functional equation

$$e^{in(\phi+\psi)} = e^{in\phi} \cdot e^{in(\psi)} \tag{4.1}$$

and

$$e^{2n\pi i} = 1.$$

In the group theoretical interpretation this means that the value of the transformed function $e^{in(\phi+\psi)}$ is equal to the original function $e^{in\phi}$ times a complex factor $e^{in\psi}$. In the three-dimensional case we have as the domain the unit sphere S^2 and the rotations $R \in SO(3)$ transform a function on the sphere by rotating the sphere first: $f^R(X) = f(R^{-1}X)$.

If we have a finite dimensional subspace of $L^2(S^2)$ that is invariant under this action of $SO(3)$ then we can select a basis $Y^1, ..., Y^m$ of this subspace and we get for the transformed functions $Y^k(R^{-1}X)$ the equation:

$$Y^k(R^{-1}X) = \sum_{l=1}^{m} t_{lk}(R)Y^l(X)$$

Writing this in matrix form gives:

$$Y(R^{-1}X) = T(R)Y(X) \tag{4.2}$$

Equations 4.1 and 4.2 describe the same type of transformation with the only difference that we have replaced the function itself with a vector of functions and the complex factor becomes a complex matrix. The functions Y_l^m have thus the same transformation property as the complex exponentials. This is the more practical side of this chapter. On the more theoretical side this chapter demonstrates some of the basic techniques from Lie theory.

In this chapter we follow closely the book of Gelfand et al. [15]. For more information on Lie theory the reader may consult one of the books on Lie theory (for example [47] or [55]).

4.1.1   Exponential of a Matrix

Let A be any $n \times n$ matrix with elements a_{ij} and assume that these elements have a common upper bound $|a_{ij}| \leq C$. Since A is a square matrix we can compute A^m for all integers m. If we denote the elements of A^m by $a_{ij}^{(m)}$ then we find with a simple computation:

$$\left| a_{ij}^{(m)} \right| \leq (nC)^m$$

From this inequality we see that the following definition is meaningful:

Definition 4.1

$$e^A = \sum_{m=0}^{\infty} \frac{1}{m!} A^m$$

We call e the *exponential function.*

For the exponential function we collect a number of useful properties in the next theorem:

Theorem 4.1   1. If B is an invertible $n \times n$ matrix then we have

$$e^{BAB^{-1}} = Be^A B^{-1}$$

2. If $\lambda_i (i = 1, \ldots n)$ are the eigenvalues of A then the eigenvalues of e^A are given by e^{λ_i}.

3. $\det(e^A) = e^{\text{tr} A}$

4. e^A is regular for all matrices A.

5. If $AB = BA$ then $e^{A+B} = e^A e^B$.

Most of these properties are simple consequences of the definition of the exponential function; the second part of the theorem about the eigenvalues is proved by induction on the number of eigenvalues (for a proof of this theorem and the next two see [5]).

Theorem 4.2 Let $t \in \mathbf{R}$ be a real variable and A be a fixed $n \times n$ matrix. The mapping $e^. \mathbf{R} \rightarrow GL(n, \mathbf{C}); t \mapsto e^{tA}$ is a continuous homomorphism of the additive group of reals into $GL(n, \mathbf{C})$.

Theorem 4.3 There is a neighborhood U of the zero matrix in the space of all complex $n \times n$ matrices that is homeomorphic to a neighborhood V of E in $GL(n, \mathbf{C})$. The homeomorphism is given by $A \mapsto e^A$.

4.1.2   Infinitesimal Rotations

In this section we investigate an irreducible, unitary representation of $SO(3)$ as a matrix function parametrized by the rotation axis and the rotation angle. We show that this highly nonlinear function can be completely described by three matrices.

Suppose that T is an irreducible, unitary representation of $SO(3)$ and that we selected a fixed orthonormal basis in the representation space. Then T is completely described by a unitary matrix. This matrix will also be denoted by T. We describe a rotation g by the vector (ξ_1, ξ_2, ξ_3) where the direction of the vector is given by the rotation axis and the length of the vector is equal to the rotation angle. The identity rotation is described by the zero vector. Then T becomes a function of the ξ_i and we have $T(0, 0, 0) = E$ where E is the identity matrix. It can be proved that T as a function of the parameters ξ_i is infinitely differentiable and we can thus develop T into a Taylor series:

$$T(\xi_1, \xi_2, \xi_3) = E + \xi_1 A_1 + \xi_2 A_2 + \xi_3 A_3 + \ldots \tag{4.3}$$

This is a matrix equation with constant matrices A_i. To see what these matrices are we consider the case of a rotation around the x-axis with rotation angle ξ. In this case we have:

$$T(\xi, 0, 0) = E + \xi A_1 + \ldots \tag{4.4}$$

and we find that

$$A_1 = \lim_{\xi \to 0} \frac{T(\xi, 0, 0) - T(0, 0, 0)}{\xi} = \frac{dT(\xi_1, \xi_2, \xi_3)}{d\xi 1} \tag{4.5}$$

These matrices are thus the partial derivatives of T at the point $(0, 0, 0)$ and we define:

Definition 4.2 The matrices A_i in equation 4.3 are called the *matrices of infinitesimal rotations about the coordinate axes.*

In the next section we will describe the possible forms of the $A_i's$ but now we will show in the next theorem that T is completely defined by these three matrices A_i.

Theorem 4.4 The matrices A_i determine completely the representation, i.e. given the A_i we can determine $T(\xi_1, \xi_2, \xi_3)$ for all ξ_i.

To see this take an arbitrary vector $(\xi_1, \ \xi_2, \ \xi_3)$ and two rotations $g(t\xi_1, \ t\xi_2, \ t\xi_3)$ and $g(s\xi_1, \ s\xi_2, \ s\xi_3)$. These are rotations with the common rotation axis given by the vector $(\xi_1, \ \xi_2, \ \xi_3)$ and rotation angles $t\sqrt{\xi_1^2 + \xi_2^2 + \xi_3^2}$ and $s\sqrt{\xi_1^2 + \xi_2^2 + \xi_3^2}$ respectively. The product of these two rotations is the rotation

$$
\begin{aligned}
g((s+t)\xi_1, \ &(s+t)\xi_2, \ (s+t)\xi_3) = \\
= \ g(t\xi_1, \ &t\xi_2, \ t\xi_3) g(s\xi_1, \ s\xi_2, \ s\xi_3)
\end{aligned}
\tag{4.6}
$$

Since T is a representation we find:

$$
\begin{aligned}
T((s+t)\xi_1, \ &(s+t)\xi_2, \ (s+t)\xi_3) = \\
= \ T(s\xi_1, \ &s\xi_2, \ s\xi_3) T(t\xi_1, \ t\xi_2, \ t\xi_3)
\end{aligned}
\tag{4.7}
$$

We differentiate both side of equation 4.7 with respect to s and set $s = 0$ to get the differential equation:

$$
\frac{d}{dt}T(t\xi_1, t\xi_2, t\xi_3) = \frac{d}{ds}T(s\xi_1, s\xi_2, s\xi_3)|_{s=0} T(t\xi_1, t\xi_2, t\xi_3)
\tag{4.8}
$$

From the Taylor expansion 4.3 we find

$$
\frac{d}{ds}T(s\xi_1, s\xi_2, s\xi_3)|_{s=0} = A_1\xi_1 + A_2\xi_2 + A_3\xi_3
\tag{4.9}
$$

If we set $X(t) = T(t\xi_1, t\xi_2, t\xi_3)$ then we get the differential equation for X:

$$
\frac{d}{dt}X(t) = (A_1\xi_1 + A_2\xi_2 + A_3\xi_3)X(t)
\tag{4.10}
$$

The solution that obeys the initial value $X(0) = E$ is given by

$$
X(t) = e^{t(A_1\xi_1 + A_2\xi_2 + A_3\xi_3)}
\tag{4.11}
$$

or

$$
T(\xi_1, \xi_2, \xi_3) = e^{A_1\xi_1 + A_2\xi_2 + A_3\xi_3} = e^{A_\xi}
\tag{4.12}
$$

where we used the abbreviation $A_\xi = A_1\xi_1 + A_2\xi_2 + A_3\xi_3$.

This shows that the representation is completely specified by the matrices A_i.

4.1.3   The Commutation Relations

In this section we investigate the relations between the infinitesimal rotations A_i.

If g_0 is a fixed rotation and g is another rotation then we find that the rotations $\tilde{g}_0 = gg_0g^{-1}$ and G_0 have the same rotation angle. This follows from the fact that $\tilde{g}_0$ and g_0 have the same eigenvalues and from the connection between the rotation angle and the eigenvalues. Now we describe the rotation g_0 by the vector η as defined in the previous section. Applying the rotation g to this vector η gives a new vector $\tilde{\eta} = g\eta$. η describes the rotation axis of g_0 and we have therefore $g_0\eta = \eta$. From this we get $\tilde{g}_0\tilde{\eta} = gg_0g^{-1}\tilde{\eta} = gg_0\eta = \tilde{\eta}$. Since g_0 and $\tilde{g}_0$ have the same rotation angle we find that $\tilde{\eta}$ describes the rotation $\tilde{g}_0$.

For notational convenience we write now T_g instead of $T(g)$. From the representation property we find:

$$T_{\tilde{g}_0} = T_{gg_0g^{-1}} = T_gT_{g_0}T_{g^{-1}}.$$

Inserting this into the Taylor series gives:

$$T_{\tilde{g}_0} = e^{A\eta} = T_{gg_0g^{-1}} = T_gT_{g_0}T_{g^{-1}} = T_ge^{A\eta}T_{g^{-1}} \tag{4.13}$$

Using theorem 4.1 we get:

$$T_gA_\eta T_{g^{-1}} = A_{\tilde{\eta}} \tag{4.14}$$

Now select as g a rotation with angle α around the x-axis, i.e. $T_g = T(\alpha,0,0) = e^{A_1\cdot\alpha}$ and $\eta = (0,\ 1,\ 0)$. Since $\tilde{\eta} = g\eta$ we have also $\tilde{\eta} = (0,\ \cos\alpha,\ \sin\alpha)$. Expanding the exponential function we find therefore from 4.14 and $T_{g^{-1}} = T(-\alpha,0,0)$:

$$A_2 + \alpha(A_1A_2 - A_2A_1) + \alpha^2... = \cos\alpha A_2 + \sin\alpha A_3 \tag{4.15}$$

and from this we get $A_1A_2 - A_2A_1 = A_3$

Definition 4.3 If A and B are two square-matrices then the expression $AB - BA$ is called the *commutator* of A and B. We write:

$$[A, B] = AB - BA \tag{4.16}$$

This expression is sometimes also called the *bracket* of A and B.

We find by similar computations the following theorem:

Theorem 4.5 If $T : g \to T(g)$ is an arbitrary representation of $SO(3)$ and A_1, A_2 and A_3 are the matrices corresponding to the infinitesimal rotations about the coordinate axes then we have:

$$\begin{aligned} [A_1, A_2] &= A_3 \\ [A_2, A_3] &= A_1 \\ [A_3, A_1] &= A_2 \end{aligned} \tag{4.17}$$

There is the following relation between the commutator and the vector product (see [15]):

Theorem 4.6 For a vector $(\xi_1,\ \xi_2,\ \xi_3)$ we define as usual the matrix $A_\xi = A_1\xi_1 + A_2\xi_2 + A_3\xi_3$ with the infinitesimal matrices A_i. Then we have the following relation between the bracket and the vector product: if ξ, η are two 3-D vectors and $\zeta = \xi \times \eta$ is the vector product of ξ and η then we have

$$[A_\xi, A_\eta] = A_\zeta \tag{4.18}$$

4.1.4   Canonical Basis of an Irreducible Representation

In the next theorem we describe what consequences the unitary constraint has on the matrices A_i. T was a unitary representation and the matrices T_g are therefore unitary matrices satisfying: $T_g^* T_g = E$. If we choose g as the rotation around the x-axis then we find: $T(\xi, 0, 0)^* T(\xi, 0, 0) = E$ (where T^* is the conjugate complex transpose of T). Inserting the Taylor expansion for T and comparing the linear entries we find

$$A_1 + A_1^* = 0 \tag{4.19}$$

or $A_1 = -A_1^*$ and therefore:

Theorem 4.7 Define $H_k = iA_k$, then we have

1. H_k are Hermitian matrices, i.e. $H_k = H_k^*$

2. The H_k satisfy the following relations:

$$
\begin{aligned}
\left[H_1, H_2\right] &= iH_3 \\
\left[H_2, H_3\right] &= iH_1 \\
\left[H_3, H_1\right] &= iH_2
\end{aligned}
\tag{4.20}
$$

In the next theorem we introduce two new matrices which will be more convenient in the following computations.

Theorem 4.8 If the operators H_+, H_- are defined as:

$$
\begin{aligned}
H_+ &= H_1 + iH_2 \\
H_- &= H_1 - iH_2
\end{aligned}
\tag{4.21}
$$

then we have:

1.

$$H_+^* = H_- \tag{4.22}$$

2. The matrices H_+, H_-, H_3 satisfy the following relations:

$$
\begin{aligned}
\left[H_+, H_3\right] &= -H_+ \\
\left[H_-, H_3\right] &= H_- \\
\left[H_+, H_-\right] &= 2H_3
\end{aligned}
\tag{4.23}
$$

We now proceed as follows: First we find all matrices H_+, H_-, H_3 that satisfy the conditions of theorem 4.8, then we use 4.21 to find the matrices H_1, H_2 and H_3. These in turn are then used in theorem 4.7 to find the matrices A_k which define the representation.

For the eigenvectors of H_3 we find the following theorem:

Theorem 4.9 Let f be an eigenvector of H_3 with eigenvalue λ:

$$H_3 f = \lambda f$$

and define:

$$
\begin{aligned}
f_+ &= H_+ f \\
f_- &= H_- f
\end{aligned}
$$

then we find:

1. f_+ is either the zero-vector or an eigenvector of H_3 with eigenvalue $\lambda + 1 : H_3 f_+ = (\lambda + 1) f_+$

2. f_- is either the zero-vector or an eigenvector of H_3 with eigenvalue $\lambda - 1 : H_3 f_- = (\lambda - 1) f_-$

Definition 4.4 The operators H_+ and H_- are called the *raising and lowering operators.*

We show only the first part concerning f_+ :

$$H_3 f_+ = H_3 H_+ f = [H_3, H_+] f + H_+ H_3 f =$$
$$H_+ f + H_+ \lambda f = (\lambda + 1) H_+ f = (\lambda + 1) f_+ \tag{4.24}$$

Now we proceed to construct H_+, H_-, H_3 : H_3 is Hermitian and its eigenvalues are therefore real. Let l be the largest eigenvalue with (normalized) eigenvector f_l. If $H_- f_l \neq 0$ then we define the unit vector f_{l-1} by $H_- f_l = \lambda_l f_{l-1}$. From the previous theorem 4.9 we find that f_{l-1} is also an eigenvector of H_3 with eigenvalue $l - 1$. Continuing this process we find a series of eigenvectors $f_l, f_{l-1}, f_{l-2}, \ldots$ of H_3 with eigenvalues $l, l - 1, l - 2, \ldots$. This process must stop after finitely many steps, say with the index $k : H_- f_k = 0$. We thus constructed a series of vectors f_m that satisfy the following conditions:

$$H_3 f_m = m f_m \tag{4.25}$$

and

$$H_- f_m = \lambda_m f_{m-1} \tag{4.26}$$

$H_+ f_m$ is either the zero vector or an eigenvector of H_3 and since f_l belongs to the largest eigenvalue we have $H_+ f_l = 0$. For $H_+ f_{l-1}$ we get:

$$H_+ f_{l-1} = \frac{1}{\lambda_l} H_+ H_- f_l = \frac{1}{\lambda_l} [H_+, H_-] f_l + \frac{1}{\lambda_l} H_- H_+ f_l = \frac{2}{\lambda_l} H_3 f_l = \frac{2l}{\lambda_l} f_l. \tag{4.27}$$

We define β_l by: $H_+ f_{l-1} = \beta_l f_l$ and we find for the other f_m by induction:

$$
\begin{aligned}
H_+ f_m &= \frac{1}{\lambda_{m+1}} H_+ H_- f_{m+1} \\
&= \frac{1}{\lambda_{m+1}} [H_+, H_-] f_{m+1} + \frac{1}{\lambda_{m+1}} H_- H_+ f_{m+1} \\
&= \frac{2}{\lambda_{m+1}} H_3 f_{m+1} + \frac{\beta_{m+2}}{\lambda_{m+1}} H_- f_{m+2}
\end{aligned}
\tag{4.28}
$$

Using equation 4.26 we get

$$H_+ f_m = \frac{2(m + 1) + \lambda_{m+2} \beta_{m+2}}{\lambda_{m+1}} f_{m+1} \tag{4.29}$$

and if we set

$$\beta_{m+1} = \frac{2(m + 1) + \lambda_{m+1} \beta_{m+2}}{\lambda_{m+1}} \tag{4.30}$$

then equation 4.29 becomes

$$H_+ f_m = \beta_{m+1} f_{m+1}. \tag{4.31}$$

Using equations 4.26 and 4.31 and $H_+^* = H_-$ from equation 4.22 we get:

$$\beta_m \langle f_m, f_m \rangle = \langle H_+ f_{m-1}, f_m \rangle = \langle f_{m-1}, H_- f_m \rangle = \lambda_m \langle f_{m-1}, f_{m-1} \rangle \tag{4.32}$$

The eigenvectors are normalized and we find $\lambda_m = \beta_m$.

Using this relation between the λ's and β's in equation 4.30 we find $\lambda_m^2 - \lambda_{m+1}^2 = 2m$ and by a summation:

$$\lambda_m^2 = \lambda_m^2 - \lambda_{l+1}^2 = 2l + 2(l-1) + \cdots + 2m = (l+m)(l-m+1) \tag{4.33}$$

Now k was the index of the lowest eigenvector and therefore $\lambda_k = 0$. Consequently $0 = \lambda_k^2 = (l+k)(l-k+1)$ and therefore $k = -l$. l was the largest eigenvalue, $-l$ the lowest and the eigenvalue was decremented by one in each processing step. It follows that the number of eigenvectors is $2l + 1$ and l is therefore an integer or half an odd integer.

Under H_+, H_- and H_3 an eigenvector f_m is carried into another eigenvector and since we assumed that the representation is irreducible we see that the constructed space of eigenvectors of H_3 is the whole representation space. We collect this result in the following theorem:

Theorem 4.10 For any irreducible representation the transformations H_+, H_- and H_3 define an orthogonal basis consisting of the normalized eigenvectors f_m of H_3. These eigenvectors satisfy the following conditions:

$$\begin{aligned}
H_+ f_m &= \lambda_{m+1} f_{m+1} \\
H_- f_m &= \lambda_m f_{m-1} \\
H_3 f_m &= m f_m.
\end{aligned} \tag{4.34}$$

We have $m = -l, -l+1, \ldots, l$ with an integer or half an odd integer l and for the constants λ_m we find: $\lambda_m = \sqrt{(l+m)(l-m+1)}$

Definition 4.5 The basis $f_{-l}, \ldots, f_l$ of the normalized eigenvectors of H_3 is called the *canonical basis of the representation* .

Returning to our initial problem of finding the possible matrices of the infinitesimal rotations A_k we have:

Theorem 4.11 Assume that $\lambda_m = \sqrt{(l+m)(l-m+1)}$, that f_m is the canonical basis and that $m = -l, -l+1, \ldots, l$. Then we have:

1. Each irreducible representation of $SO(3)$ is defined by an integer or half an odd integer l. The matrices A_k are given by the conditions:

$$\begin{aligned}
A_1 f_m = -i H_1 f_m &= -\frac{i}{2}(H_+ + H_-) f_m = -\frac{i}{2}\lambda_{m+1} f_{m+1} - \frac{i}{2}\lambda_m f_{m-1} \\
A_2 f_m = -i H_2 f_m &= -\frac{1}{2}(H_+ - H_-) f_m = -\frac{1}{2}\lambda_{m+1} f_{m+1} + \frac{1}{2}\lambda_m f_{m-1} \\
A_3 f_m = -i H_3 f_m &= -i m f_m
\end{aligned} \tag{4.35}$$

2. The matrices have the following form:

$$
A_1 = -\frac{i}{2}
\begin{pmatrix}
0 & \lambda_{-l+1} & 0 & \cdots & 0 & 0 \\
\lambda_{-l+1} & 0 & \lambda_{-l+2} & \cdots & 0 & 0 \\
\cdots & \cdots & \cdots & \cdots & \cdots & \cdots \\
0 & 0 & 0 & \cdots & 0 & \lambda_l \\
0 & 0 & 0 & \cdots & \lambda_l & 0
\end{pmatrix}
$$

$$
A_2 = \frac{1}{2}
\begin{pmatrix}
0 & \lambda_{-l+1} & 0 & \cdots & 0 & 0 \\
-\lambda_{-l+1} & 0 & \lambda_{-l+2} & \cdots & 0 & 0 \\
\cdots & \cdots & \cdots & \cdots & \cdots & \cdots \\
0 & 0 & 0 & \cdots & 0 & \lambda_l \\
0 & 0 & 0 & \cdots & -\lambda_l & 0
\end{pmatrix}
$$

$$
A_3 =
\begin{pmatrix}
il & 0 & 0 & \cdots & 0 & 0 \\
0 & i(l-1) & 0 & \cdots & 0 & 0 \\
\cdots & \cdots & \cdots & \cdots & \cdots & \cdots \\
0 & 0 & 0 & \cdots & -i(l-1) & 0 \\
0 & 0 & 0 & \cdots & 0 & -il
\end{pmatrix}
\tag{4.36}
$$

Definition 4.6 Each irreducible representation of $SO(3)$ is uniquely defined by the number l given in the last theorem 4.11. l is called the *weight* of the representation.

Up to now we have shown that for an irreducible representation T the matrices T_g are of the form $T_g = T(\xi_1, \xi_2, \xi_3) = T(\xi) = e^{A\xi}$ (see equation 4.12) and that the matrices A_k are given by the conditions in theorem 4.11. However, we have not shown that there is such a representation for each l. We will later construct such a representation. Now we will only show that if there is a representation of the form given in equation 4.12 with the A_k satisfying the conditions in theorem 4.11 then this representation is irreducible.

We show that every invariant subspace has dimension $2l + 1$. Consider an invariant subspace and denote the largest eigenvector of H_3 in this subspace by $h = \sum_{m=-l}^{l} c_m f_m$. From theorem 4.9 we find that $H_+ h$ is the zero vector and from theorem 4.10 we get

$$
0 = H_+ h = \sum_{m=-l}^{l} c_m H_+ f_m = \sum_{m=-l}^{l} c_m \lambda_{m+1} f_m
$$

The f_m are independent and therefore $c_m \lambda_{m+1} = 0$ for all m. If $m < l$ then $\lambda_{m+1} \neq 0$ and therefore $c_m = 0$ and from this we see that f_l is an element of the invariant subspace. Since the space is invariant under H_- we find that also all the other f_m are elements of the invariant subspace.

In the next theorem we introduce still another transformation and we show how it can be used to characterize the representation space.

Theorem 4.12 Define the transformation H^2 as

$$
H^2 = H_1^2 + H_2^2 + H_3^2
\tag{4.37}
$$

Then we have:

1. $[H^2, H_1] = [H^2, H_2] = [H^2, H_3]$

2. All vectors f in the representation space of a representation of weight l satisfy the equation:

$$H^2 f = l(l+1)f \tag{4.38}$$

The first part of the theorem is an easy calculation using the definitions of the operators involved and the properties of the bracket. To see the second part check first that $H_+ H_- = H_1^2 + H_2^2 + H_3$ and find that $H^2 = H_+ H_- - H_3 + H_3^2$ then from theorem 4.10 deduce

$$H_+ H_- f_m = \lambda_m^2 f_m$$

$$H_3 f_m = m f_m$$

$$H_3^2 f_m = m^2 f_m$$

But $\lambda_m = \sqrt{(l+m)(l-m+1)}$ and therefore we get:

$$H^2 f_m = (\lambda_m^2 - m + m^2) f_m = l(l+1) f_m$$

4.1.5   Spherical Functions

In the last section we saw how to characterize a given irreducible representation. In this section we will now actually construct these representations. For this purpose we consider the function space L of differentiable functions on S^2, the unit sphere in three-dimensional space. On S^2 we introduce polar coordinates φ, θ defined by the equations:

$$
\begin{aligned}
x &= \cos \varphi \sin \theta \\
y &= \sin \varphi \sin \theta \\
z &= \cos \theta.
\end{aligned}
\tag{4.39}
$$

or

$$
\begin{aligned}
\varphi &= \arctan \frac{y}{x} \\
\theta &= \arctan \frac{\sqrt{x^2 + y^2}}{z}.
\end{aligned}
\tag{4.40}
$$

The scalar product in this space is given by

$$\langle f, g \rangle = \int_0^{2\pi} \int_0^{\pi} f(\theta, \varphi) \overline{g(\theta, \varphi)} \sin \theta \, d\theta d\varphi = \int_{S^2} f \bar{g} \, d\omega \tag{4.41}$$

As usual we define the representation T by $T(g)f(X) = f(g^{-1}(X))$. This is indeed a representation as can be seen by checking that $T(g)f$ is indeed differentiable. We define

Definition 4.7   1. The functions on the sphere that belong to an irreducible representation of weight l are called the *spherical functions of order l.*

2. The functions $f_m(\theta, \varphi)$ forming the canonical basis in the space of spherical functions of order l are called the *basic spherical functions of the l-th order.*

3. The basic spherical functions of the l-th order will be denoted by $Y_l^m(\theta, \varphi)$

We will now derive these basic functions Y_l^m. In the last sections we saw that the canonical basis was defined with the help of the matrices A_k of the infinitesimal rotations and we must therefore first obtain the transformations which correspond to these infinitesimal rotations.

We first consider A_3 describing an infinitesimal rotation around the z-axis:

Theorem 4.13

$$A_3 f = -\frac{\partial f(\theta, \varphi)}{\partial \varphi}$$

To see this take a rotation g around the z-axis with rotation angle α. Then we have $T(g) = e^{\alpha A_3}$ and we expand therefore $T(g)f$ in terms of α:

$$T(g)f(\theta, \varphi) = f(g^{-1}(\theta, \varphi)) = f(\theta, \varphi - \alpha)$$

and from

$$f(\theta, \varphi - \alpha) = f(\theta, \varphi) - \alpha \frac{\partial f(\theta, \varphi)}{\partial \varphi} + \ldots \tag{4.42}$$

we find the form of A_3. For the matrices A_1 and A_2, i.e. the rotations around the x- and y-axis we get

Theorem 4.14

$$
\begin{aligned}
A_1 &= \sin \varphi \frac{\partial}{\partial \theta} + \cot \theta \cos \varphi \frac{\partial}{\partial \varphi} \\
A_2 &= -\cos \varphi \frac{\partial}{\partial \theta} + \cot \theta \sin \varphi \frac{\partial}{\partial \varphi}
\end{aligned}
\tag{4.43}
$$

We compute only the expression for the x-axis rotation, the other expression can be derived similarly.

If g is a rotation around the x-axis with rotation angle then we expand again $T(g)f$ in a series in α and we get then A_1 as the operator belonging to the linear term in α. We thus get:

$$A_1 = \lim_{\alpha \to 0} \frac{T(\alpha, 0, 0) - T(0, 0, 0)}{\alpha}$$

If we expand $T(\alpha, 0, 0)f$ into a Taylor series then we get

$$T(\alpha, 0, 0)f(\theta, \varphi) = f(\theta, \varphi) + \left(\frac{\partial f}{\partial \theta} \frac{\partial \theta}{\partial \alpha} + \frac{\partial f}{\partial \varphi} \frac{\partial \varphi}{\partial \alpha} \right) |_{\alpha=0} \alpha + \ldots$$

If (x', y', z') is the rotated coordinate system $g^{-1}(x, y, z)$ then we see that

$$
\begin{aligned}
\frac{\partial x'}{\partial \alpha} &= 0 \\
\frac{\partial y'}{\partial \alpha} |_{\alpha=0} &= z \\
\frac{\partial z'}{\partial \alpha} |_{\alpha=0} &= -y
\end{aligned}
\tag{4.44}
$$

But

$$
\begin{aligned}
\frac{\partial x'}{\partial \alpha} &= \frac{\partial x'}{\partial \theta}\frac{\partial \theta}{\partial \alpha} + \frac{\partial x'}{\partial \varphi}\frac{\partial \varphi}{\partial \alpha} \\
&= \cos\theta\cos\varphi\frac{\partial \theta}{\partial \alpha} - \sin\theta\sin\varphi\frac{\partial \varphi}{\partial \alpha}
\end{aligned}
\tag{4.45}
$$

leading to

$$
0 = \cos\theta\cos\varphi\frac{\partial \theta}{\partial \alpha} - \sin\theta\sin\varphi\frac{\partial \varphi}{\partial \alpha}
\tag{4.46}
$$

For y' and z' we get with the same calculations:

$$
\begin{aligned}
\cos\theta &= \cos\theta\sin\varphi\frac{\partial \theta}{\partial \alpha} + \sin\theta\cos\varphi\frac{\partial \varphi}{\partial \alpha} \\
-\sin\theta\sin\varphi &= -\sin\theta\frac{\partial \theta}{\partial \alpha}
\end{aligned}
\tag{4.47}
$$

From equations 4.46 and 4.47 we find

$$
\begin{aligned}
\frac{\partial \theta}{\partial \alpha} &= \sin\varphi \\
\frac{\partial \varphi}{\partial \alpha} &= \cot\theta\cos\varphi
\end{aligned}
\tag{4.48}
$$

Inserting these expressions in the Taylor series for $T(g)f$ we get the following expression for A_1:

$$
T(g)f = f(\theta,\varphi) + \left(\frac{\partial f}{\partial \theta}\sin\varphi + \frac{\partial f}{\partial \varphi}\cot\theta\cos\varphi\right)|_{\alpha=0}\,\alpha + \ldots
$$

From the expressions for the operators A_k it is now easy to compute the differential operators belonging to the operators H_3, H_+, and H_-.

Theorem 4.15

$$
\begin{aligned}
H_+ &= e^{i\varphi}\left(\frac{\partial}{\partial \theta} + i\cot\theta\frac{\partial}{\partial \varphi}\right) \\
H_- &= e^{-i\varphi}\left(-\frac{\partial}{\partial \theta} + i\cot\theta\frac{\partial}{\partial \varphi}\right) \\
H_3 &= -i\frac{\partial}{\partial \varphi}
\end{aligned}
\tag{4.49}
$$

From the previous derivations we know that the basic spherical functions Y_l^m are given by the eigenfunctions of the H_3 operator, i.e. they are the solutions of the differential equation:

$$
H_3 Y_l^m(\theta,\varphi) = -i\frac{\partial}{\partial \varphi}Y_l^m(\theta,\varphi) = mY_l^m(\theta,\varphi)
$$

Hence we find:

$$
Y_l^m(\theta,\varphi) = \frac{1}{\sqrt{2\pi}}e^{im\varphi}F_l^m(\theta)
\tag{4.50}
$$

Since Y is a one-valued function on the sphere and since $Y_l^m(\theta, \varphi) = Y_l^m(\theta, \varphi + 2\pi)$ we find that m is an integer. Since the Y are also normalized we get from equation 4.50 the following integral for the function F:

$$1 = \int_0^\pi |F_l^m(\theta)|^2 \sin\theta \, d\theta \qquad (4.51)$$

Our next task is the derivation of the function F. This will be done with the help of the H^2 operator. Recall (theorem 4.12) that all functions in the representation space satisfy the equation $H^2 f = l(l+1)f$. Using theorem 4.15 we find for H^2 the *differential equation of the spherical functions of the l-th order*:

$$-H^2 = \frac{1}{\sin\theta} \frac{\partial}{\partial\theta} \left(\sin\theta \frac{\partial}{\partial\theta} \right) + \frac{1}{\sin^2\theta} \frac{\partial^2}{\partial\varphi^2} \qquad (4.52)$$

Substituting the expression for Y_l^m from equation 4.50 into 4.52 we find the differential equation for F_l^m:

$$\frac{1}{\sin\theta} \frac{\partial}{\partial\theta} \left(\sin\theta \frac{\partial F_l^m}{\partial\theta} \right) + \left[l(l+1) - \frac{m^2}{\sin^2\theta} \right] F_l^m(\theta) = 0 \qquad (4.53)$$

or after the substitution $\mu = \cos\theta$, $P_l^m(\mu) = F_l^m(\cos\theta)$:

$$\frac{\partial}{\partial\mu} \left[(1-\mu^2) \frac{\partial P_l^m(\mu)}{\partial\mu} \right] + \left[l(l+1) - \frac{m^2}{1-\mu^2} \right] P_l^m(\mu) = 0 \qquad (4.54)$$

We summarize this in the next theorem:

Theorem 4.16 The basic spherical functions have the form:

$$Y_l^m(\theta, \varphi) = \frac{1}{\sqrt{2\pi}} e^{im\varphi} P_l^m(\cos\theta)$$

where the P_l^m are solutions of the differential equation 4.54.

It now remains to find an expression for the dependence of Y_l^m on θ. The next theorem, together with theorem 4.16 will completely specify the form of the basis functions:

Theorem 4.17   1. The functions $P_l^m(\cos\theta)$ defined as in theorem 4.16 are given by:

$$P_l^m(\mu) = \sqrt{\frac{(l+m)!}{(l-m)!}} \sqrt{\frac{2l+1}{2}} \frac{1}{2^l \cdot l!} \left(1-\mu^2\right)^{-m/2} \frac{d^{l-m}(\mu^2-1)^l}{d\mu^{l-m}} \qquad (4.55)$$

2. If we write $P_l(\mu) = P_l^0(\mu)$ then we have:

$$P_l(\mu) = \sqrt{\frac{2l+1}{2}} \frac{1}{2^l \cdot l!} \frac{d^l(\mu^2-1)^l}{d\mu^l} \qquad (4.56)$$

Definition 4.8   1. The polynomial P_l is called the *normalized Legendre polynomial of order l*.

2. The polynomial P_l^m is called the *normalized associated Legendre function.*

We consider first Y_l^l the solution of the equations:

$$\begin{aligned} H_3 Y_l^l &= l Y_l^l \\ H_+ Y_l^l &= 0 \end{aligned} \tag{4.57}$$

We saw in equation 4.50 that the function $Y_l^l(\theta, \varphi)$ is of the form $Y_l^l(\theta, \varphi) = \frac{1}{\sqrt{2\pi}} e^{il\varphi} F_l^l(\theta)$. From the form of H_+ (theorem 4.15) and the second equation in 4.57 we get:

$$\frac{dF_l^l(\theta)}{d\theta} - l \cot \theta F_l^l(\theta) = 0 \tag{4.58}$$

The general solution of this differential equation is given by

$$F_l^l(\theta) = C \sin^l \theta \tag{4.59}$$

Using the normalizing condition in equation 4.51 we find for the constant C the value

$$C = (-1)^l \frac{1}{2^l l!} \sqrt{\frac{2l+1}{2}} \sqrt{(2l)!} \tag{4.60}$$

Having found the form of Y_l^l we use the equation $H_- Y_l^m = \lambda_m Y_l^{m-1}$ to get the expression for the other functions. Using the expression form H_- we get

$$e^{-i\varphi} \left(-\frac{\partial Y_l^m}{\partial \theta} + i \cot \theta \frac{\partial Y_l^m}{\partial \varphi} \right) = \lambda_m Y_l^{m-1} \tag{4.61}$$

Inserting 4.50 and canceling factors depending only on φ we get the following recurrence relation for F_l^m :

$$-\frac{dF_l^m(\theta)}{d\theta} - m \cot \theta F_l^m(\theta) = \lambda_m F_l^{m-1}(\theta) \tag{4.62}$$

Making again the substitution $\mu = \cos \theta$ and $P_l^m(\mu) = F_l^m(\theta)$ gives:

$$\sqrt{1/\mu^2} \left(\frac{dP_l^m(\mu)}{\mu} - m \frac{\mu}{1-\mu^2} P_l^m(\mu) \right) = \lambda_m P_l^{m-1}(\mu) \tag{4.63}$$

Using this recurrence relation between the P_l^m it is possible to derive the required expression for the functions P_l^m.

4.1.6   Homogeneous Polynomials

Before we leave the group $SO(3)$ we will present a more traditional approach to surface harmonics. We first derive a transformation formula that describes explicitly how the surface harmonics transform under a rotation. This is important in the pattern recognition applications of representation theory for the following reason.

Assume we have a fixed pattern $p_0(x, y, z)$ like an edge or a line in 3-D. We approximate this function by a sum $p_o \approx \sum_{l=0}^{L} \sum_{m=-l}^{l} a_{lm}(p_0) Y_l^m$ and describe the function p_0 by the measurements $\{a_{lm}(p_0)\}$. For ease of notation we collect the coefficients $a_{lm}(p_0)$ belonging to the same l in the vector: $A_l(p_0) = (a_{-ll} \quad \ldots \quad a_{ll})$.

Now we know that if another pattern p is a rotated version of p_0 (i.e. $p = p_0^R$ for some $R \in SO(3)$) then we can find unitary matrices $T_l(R)$ such that $A_l(p) = A_l(p_0^R) = T_l(R)A_l(p_0)$. The spatial relation between the pattern $p = p_0^R$ and the pattern p_0, given by the rotation R, is thus encoded in the transformation matrices $T_l(R)$. In our pattern recognition application we now have the two vectors $A_l(p)$ and $A_l(p_0)$. $A_l(p)$ was computed in the given image and $A_l(p_0)$ is given by our knowledge of the pattern p_0. Given these two vectors we would like to compute the orientation of the pattern p with respect to the given pattern p_0. In the edge- or line-detection example we would like to have an estimation of the orientation of the edge or line at a given point in the image. We must thus compute the rotation R from the knowledge of the vectors $A_l(p)$ and $A_l(p_0)$ and the relation $A_l(p) = T_l(R)A_l(p_0)$. An explicit knowledge of the transformation matrices $T_l(R)$ is thus of great practical importance.

The following derivation of these transformation matrices is based on the traditional treatment of the theory of special functions like the one found in [11].

There one starts with homogeneous, harmonic polynomials defined as follows:

Definition 4.9   1. Let $p \geq 0$ and $H_n(\xi_1, ..., \xi_m)$ be a polynomial of degree n in the m variables ξ_k. H_n is called a *homogeneous* of degree n if:

$$H_n(\lambda \cdot \xi_1, ..., \lambda \cdot \xi_m) = \lambda^n \cdot H_n(\xi_1, ..., \xi_m) \tag{4.64}$$

2. A function f in the m variables ξ_k is called a *harmonic* function if it is a solution to the differential equation:

$$\Lambda^m f = 0 \tag{4.65}$$

where

$$\Lambda^m = \frac{\partial^2}{\partial \xi_1^2} + \ldots + \frac{\partial^2}{\partial \xi_m^2} \tag{4.66}$$

is the Laplacian in m-dimensional space.

Starting from these definitions one can now introduce surface harmonics as follows:

Definition 4.10 Assume H_n is a harmonic, homogeneous polynomial of degree n in the m variables ξ_k. Then we call the function

$$r^{-n} H_n(\xi_1, ..., \xi_m) = H_n(\xi_1/r, ..., \xi_m/r) \tag{4.67}$$

with $r^2 = \xi_1^2 + ... + \xi_m^2$ a *surface harmonic* of degree n.

Surface harmonics are functions defined on the surface of the sphere and the space of all surface harmonics of a fixed degree is invariant under rotations of the sphere. This space of surface harmonics of a fixed degree defines thus a representation space for the group $SO(m)$. For 3-D rotations we find thus again the functions Y_l^m from equation 4.16. For a fixed degree n this space has the dimension $2n + 1$. In the general case we find the following formula for the number of orthogonal surface harmonics of a fixed degree:

Theorem 4.18 If $h(n,p)$ denotes the number of linearly independent surface harmonics of degree n with the $(p+2)$ variables $\xi_1, ..., \xi_{p+2}$ then we have:

$$h(n,p) = (2n+p)\frac{(n+p-1)!}{p!n!} \tag{4.68}$$

In the special cases of 2-D, 3-D and 4-D surface harmonics we find:

$$h(n,0) = 2, \quad h(n,1) = (2n+1), \quad h(n,2) = (n+1)^2 \tag{4.69}$$

For the 4-D surface harmonics we find the following expression (see [11]):

Theorem 4.19 Let $\eta = (\eta_1, \quad \eta_2, \quad \eta_3, \quad \eta_4)$ be a four-dimensional unit vector and denote the $(n+1)^2$ surface harmonics of degree n by $S_n^{(k,l)}(0 \le k,l \le n)$. Then we have:

1.

$$S_n^{(k,l)}(\eta) = (-1)^k(\eta_4 + i\eta_1)^{n-k-l}(\eta_3 + i\eta_2)^{k-l}P_l^{(n-k-l,k-l)}(\eta_2^2 + \eta_3^2 - \eta_1^2 - \eta_4^2) \tag{4.70}$$

if $n \ge k+l$ and

2.

$$S_n^{(k,l)}(\eta) = (-1)^{n-l}(\eta_4 - i\eta_1)^{k+l-n}(\eta_3 - i\eta_2)^{l-k}P_{n-l}^{(k+l-n,l-k)}(\eta_2^2 + \eta_3^2 - \eta_1^2 - \eta_4^2) \tag{4.71}$$

if $n < k+l$.

where $P_n^{(\alpha,\beta)}(x)$ is the *Jacobi polynomial* defined as:

$$P_n^{(\alpha,\beta)}(x) = 2^{-n}\sum_{m=0}^{n}\binom{n+\alpha}{m}\binom{n+\beta}{n-m}(x-1)^{n-m}(x+1)^m \tag{4.72}$$

Equipped with these notations we can now describe the transformation formula for the surface harmonics:

Theorem 4.20 Assume $R \in SO(3)$ is a rotation with rotation axis described by the unit vector $(x_1, \quad x_2, \quad x_3)$ and a rotation angle θ with $0 \le \theta \le \pi$. We then describe R by the vector

$$\eta = \eta(R) = (x_1\sin\tfrac{\theta}{2}, \quad x_2\sin\tfrac{\theta}{2}, \quad x_3\sin\tfrac{\theta}{2}, \quad \cos\tfrac{\theta}{2}).$$

For the surface harmonics we get the following transformation formula:

$$Y_n^k(R(\xi)) = \sum_{l=-n}^{n}(-1)^{k+l}\frac{\binom{2n}{n+k}}{\binom{2n}{n+l}}S_{2n}^{(n+k,n+l)}(\eta)Y_n^l(\xi) \tag{4.73}$$

4.2 Rigid Motions

In the last section we derived in detail the representations of the group of 3-D rotations. In this section we will briefly consider the group of rigid motions in 2-D and 3-D. We recall first the definition of a rigid motion:

Definition 4.11 A *rigid motion* in the real n-dimensional space $\mathbf{R}^n$ is a mapping of the form $X \mapsto RX + T$, where $R \in SO(n)$ is an n-dimensional rotation and T is an n-dimensional translation vector.

We note that:

1. A *rigid motion in* $\mathbf{R}^2$ has the form:

$$\begin{pmatrix} x \\ y \end{pmatrix} \mapsto \begin{pmatrix} x \cos\varphi - y \sin\varphi + \xi \\ x \sin\varphi + y \cos\varphi + \eta \end{pmatrix} \tag{4.74}$$

2. A *rigid motion in* $\mathbf{R}^2$ can thus be described by a matrix

$$\begin{pmatrix} \cos\varphi & -\sin\varphi & \xi \\ \sin\varphi & \cos\varphi & \eta \\ 0 & 0 & 1 \end{pmatrix} \tag{4.75}$$

3. The group of rigid motions in n-dimensional space is denoted by $M(n)$.

4. An *n*-dimensional rigid motion can be described by an $(n+1) \times (n+1)$-matrix of the type:

$$\begin{pmatrix} R & T \\ 0 & 1 \end{pmatrix} \tag{4.76}$$

where R is an element of $SO(n)$ and T is a translation vector from $\mathbf{R}^n$.

As representations space for $M(2)$ we take the space of all infinitely differentiable functions defined on the plane and we define for each motion g the mapping $T(g)$ as:

$$T(g)f(x,y) = f(x \cos\varphi + y \sin\varphi - \xi, -x \sin\varphi + y \cos\varphi - \eta) \tag{4.77}$$

where φ, ξ, η are the parameters describing the motion g.

The infinitesimal operators L_1, L_2, L_3 are defined as:

$$(L_1 f)(x,y) = \frac{\partial}{\partial \xi}(T(g)f)(x,y)|_{\xi=\eta=\varphi=0}$$

$$(L_2 f)(x,y) = \frac{\partial}{\partial \eta}(T(g)f)(x,y)|_{\xi=\eta=\varphi=0}$$

$$(L_3 f)(x,y) = \frac{\partial}{\partial \varphi}(T(g)f)(x,y)|_{\xi=\eta=\varphi=0} \tag{4.78}$$

and from the definition of the mapping $T(g)$ in equation 4.77 we find

$$L_1 = -\frac{\partial}{\partial x}$$

$$L_2 = -\frac{\partial}{\partial y}$$
$$L_3 = y\frac{\partial}{\partial x} - x\frac{\partial}{\partial y} \tag{4.79}$$

We introduce the new operators L_- and L_+ as

$$L_- = L_1 - iL_2$$
$$L_+ = L_1 + iL_2 \tag{4.80}$$

and find that L_3, L_+ and L_- act as the following differential operators:

$$L_3 = -\frac{\partial}{\partial \varphi}$$
$$L_- = e^{-i\varphi}\left(\frac{\partial}{\partial r} - \frac{i}{r}\frac{\partial}{\partial \varphi}\right)$$
$$L_+ = e^{i\varphi}\left(\frac{\partial}{\partial r} + \frac{i}{r}\frac{\partial}{\partial \varphi}\right). \tag{4.81}$$

where (r,φ) are the polar coordinates in the plane. Furthermore we have:

$$L_+L_- = L_-L_+ = \Lambda^2 \tag{4.82}$$

where Λ^2 is the 2-D Laplacian.

We will now find the smallest subspaces which are invariant under $M(2)$. Invariant subspaces are invariant under the operators L_i and we construct invariant subspaces for these operators.

We develop an arbitrary function f on the plane into a Fourier series in polar coordinates:

$$f(x,y) = f(r,\varphi) = \sum_m i^{-m} e^{im\varphi} g_m(r).$$

The space of functions of the form

$$\psi_m(r,\varphi) = i^{-m} e^{im\varphi} g_m(r) \tag{4.83}$$

is obviously invariant under the action of L_3.

We now start from one fixed element ψ_m and we want to find out what functions we must add to get a function space that is invariant under L_+ and L_- too. From the equations 4.81 we see that

$$L_+\psi_m = e^{i(m+1)\varphi}h(r)$$
$$L_-L_+\psi_m = e^{im\varphi}\tilde{g}(r) \tag{4.84}$$

Now we investigate if we can choose g_m in such a way that the L_+, L_- operators do not leave the invariant subspace. We thus want to choose ψ_m such that

$$L_+\psi_m = i\alpha_m\psi_{m+1}$$
$$L_-\psi_{m+1} = i\alpha_m\psi_m$$
$$L_-L_+\psi_m = -\alpha_m^2\psi_m \tag{4.85}$$

The operators L_- and L_+ commute and we find therefore $\alpha_m = \alpha_{m+1}$ or $\alpha_m = \alpha$ for all m. Therefore we find that the ψ_m must be solutions of the differential equation:

$$\Lambda^2 \psi_m = -\lambda^2 \psi_m \tag{4.86}$$

Using the Laplacian in polar coordinates we find that equation 4.86 leads to the following differential equation for the radial functions $g_m(r)$:

$$g_m''(r) - \frac{m^2}{r^2} g_m(r) + \frac{1}{r} g_m'(r) = -\lambda^2 g_m(r) \tag{4.87}$$

From the theory of Bessel functions we find therefore that $g_m(r)$ must be proportional to the Bessel function $J_m(\lambda r)$. This leads to the following theorem:

Theorem 4.21 Define X_λ is the space of all infinitely differentiable functions on the plane that solve the equation 4.86. Each non-zero value of λ leads to an irreducible representation of $M(2)$.

It can be shown that we can choose $|\lambda| = 1$ and that λ and $-\lambda$ define the same subspace. We can therefore select $\lambda = e^{i\beta}$ with $0 \leq \beta < \pi$.

The invariant subspaces under $M(2)$ are thus the subspaces of the functions of the form

$$\psi_m(x, y) = i^{-m} e^{im\varphi} J_m(\lambda r) \tag{4.88}$$

The equations 4.85 become now

$$
\begin{aligned}
\left(\frac{d}{dz} - \frac{m}{z}\right) J_m(z) &= -J_{m+1}(z) \\
\left(\frac{d}{dz} + \frac{m+1}{z}\right) J_{m+1}(z) &= J_m(z)
\end{aligned}
\tag{4.89}
$$

and a combination of these two equations results in Bessel's differential equation:

$$\left(\frac{d^2}{dz^2} + \frac{1}{z}\frac{d}{dz} + 1 - \frac{m^2}{z^2}\right) J_m(z) = 0 \tag{4.90}$$

A complete investigation of the representations of $M(2)$ can be found in [45], chapter IV.

We saw that the invariant subspaces belonging to the 2-D motion group $M(2)$ can be described by the functions $\psi_m(x, y) = i^{-m} e^{im\varphi} J_m(\lambda r)$. In a similar way it is possible to show that the representations of the 3-D motion group $M(3)$ are connected to the functions

$$Y_l^m(\theta, \varphi) J_{l+1/2}(\lambda r) \tag{4.91}$$

where $l = 0, 1, \ldots$ and $m = -l, -l+1, \ldots, l-1, l$.

For an investigation of the representations of the group $M(n)$ of motions in $\mathbf{R}^n$ the reader is referred to [48].

4.3　SL(2,C)

The special linear group $SL(2,\mathbf{C})$ is the group of all 2×2 matrices with complex entries and determinant equal to one. We write an element in $SL(2,\mathbf{C})$ as $\begin{pmatrix} a & c \\ b & d \end{pmatrix}, ac - bd = 1$. If we write an element in $\mathbf{C}^2$ as a vector $(\ z_1 \quad z_2\ )$ then we can define the following action of SL(2,C) on $\mathbf{C}^2$:

$$(\ z_1 \quad z_2\ ) \mapsto (\ z_1 \quad z_2\ ) \begin{pmatrix} a & c \\ b & d \end{pmatrix} = (\ az_1 + bz_2 \quad cz_1 + dz_2\ ) \tag{4.92}$$

Denoting the vector $(\ z_1 \quad z_2\ )$ by z and the matrix $\begin{pmatrix} a & c \\ b & d \end{pmatrix}$ by g, equation 4.92 becomes $z \mapsto zg$, where zg is the usual product of a vector and a matrix. From the last equation we find immediately that $z(g_1 g_2) = (zg_1)g_2$. It is thus easy to see that $\mathbf{C}^2$ is a homogeneous space of SL(2,C).

We now construct finite-dimensional representations of SL(2,C) in the usual way: we first select a suitable space of functions on $\mathbf{C}^2$. If f is a function in this space then we define the transformed function f^g as $f^g(z) = f(zg)$. The mapping T defined as

$$T(g)f(z) = f^g(z) = f(zg) \tag{4.93}$$

defines a representation of SL(2,C). From this (infinite-dimensional) representation we construct finite-dimensional representations by restricting T to finite-dimensional, invariant subspaces of the original function space.

In the following we consider functions $\varphi(z_1, z_2, \overline{z_1}, \overline{z_2})$ of the two complex variables z_1 and z_2. We say that φ is homogeneous of degree (n_1, n_2) if

$$\varphi(\gamma z_1, \gamma z_2, \overline{\gamma z_1}, \overline{\gamma z_2}) = \gamma^{n_1} \overline{\gamma}^{n_2} \varphi(z_1, z_2, \overline{z_1}, \overline{z_2}) \tag{4.94}$$

for all complex constants $\gamma \in \mathbf{C}$. The parameters n_1 and n_2 are complex numbers such that the difference $n_1 - n_2$ is an integer. We require this difference to be an integer since we would like $\gamma^{n_1} \overline{\gamma}^{n_2}$ to be a single valued function of γ. We will denote the degree (n_1, n_2) with the symbol χ and we will always assume that the difference between n_1 and n_2 is an integer. In the rest of this section we will write $\varphi(z_1, z_2)$ instead of $\varphi(z_1, z_2, \overline{z_1}, \overline{z_2})$.

For every degree $\chi = (n_1, n_2)$ we define the function space H_χ as the space of functions that satisfy the following two conditions:

- φ is defined for all $(z_1, z_2) \neq (0, 0)$

- $\varphi(z_1, z_2)$ is homogeneous of degree $\chi = (n_1 - 1, n_2 - 1)$

- $\varphi(z_1, z_2)$ is infinitely often differentiable in the variables z_1, z_2 and their complex conjugates.

For every element $\varphi \in H_\chi$ we introduce a new variable z as $z = z_1/z_2$ and we define a new function f by the equations

$$\varphi(z_1, z_2) = z_2^{n_1-1}\overline{z_2}^{n_2-1}\varphi(\frac{z_1}{z_2}, 1) = z_2^{n_1-1}\overline{z_2}^{n_2-1}f(\frac{z_1}{z_2}) = z_2^{n_1-1}\overline{z_2}^{n_2-1}f(z) \tag{4.95}$$

This is a function of the one extended complex variable z and the value of $f(z) = f(z_1/z_2)$ is obviously only a function of the quotient z_1/z_2. The space of these new functions will be denoted by D_χ :

$$D_\chi = \left\{ f: \text{ there is a } \varphi \in H_\chi \text{ such that: } \varphi(z_1, z_2) = z_2^{n_1-1}\overline{z_2}^{n_2-1}f(\tfrac{z_1}{z_2}) \right\} \qquad (4.96)$$

For a function $\varphi \in H_\chi$ and a matrix $g = \begin{pmatrix} a & c \\ b & d \end{pmatrix} \in SL(2,C)$ we define the following transformed function:

$$\varphi^g(z_1, z_2) = \varphi(az_1 + bz_2, cz_1 + dz_2) \qquad (4.97)$$

The transformed function φ^g is also in H_χ and we get therefore for each $g \in SL(2,C)$ a linear mapping $T_\chi(g)$ from H_χ to H_χ defined as:

$$(T_\chi(g)\varphi)(z_1, z_2) = \varphi^g(z_1, z_2) \qquad (4.98)$$

For two elements $g_1, g_2 \in SL(2,C)$ we find $T_\chi(g_1g_2) = T_\chi(g_1)T_\chi(g_2)$. The map $g \mapsto T_\chi(g)$ is thus a representation of $SL(2,C)$. The representation space is in this case H_χ. Using the connection between H_χ and D_χ described in the equations 4.95 we see that we can also use D_χ as representation space. In D_χ we get the transformation

$$f^g(z) = (cz + d)^{n_1-1}\overline{(cz + d)}^{n_2-1}f\left(\tfrac{az+b}{cz+d}\right) \qquad (4.99)$$

The map $f \mapsto f^g$ will again be denoted by $T_\chi(g)$:

$$(T_\chi(g)f)(z) = f^g(z). \qquad (4.100)$$

The spaces D_χ are thus invariant subspaces under the special linear group $SL(2,C)$.

These spaces are usually infinite dimensional but finite dimensional subspaces can be obtained for positive integers n_1 and n_2. We consider the subspaces of H_χ spanned by the polynomials:

$$P_{n_1n_2}^{\nu\mu}(z_1, z_2) = z_1^{n_1-1-\nu}z_2^{\nu}\overline{z_1}^{n_2-1-\mu}\overline{z_2}^{\mu}(0 \le \nu < n_1; 0 \le \mu < n_2) \qquad (4.101)$$

The space spanned by these polynomials has the dimension n_1n_2.

If $P_{n_1n_2}^{\nu\mu}(z_1, z_2)$ is defined as in equation 4.101 then we find the corresponding function in D_χ as

$$\begin{aligned} P(z) &= z_2^{1-n_1}\overline{z_2}^{1-n_2}P_{n_1n_2}^{\nu\mu}(z_1, z_2) \\ &= P_{n_1n_2}^{\nu\mu}(z_1/z_2, 1) = P_{n_1n_2}^{\nu\mu}(z, 1) \\ &= z^{n_1-1-\nu}\overline{z}^{n_2-1-\mu} \end{aligned}$$

We denote this polynomial again by

$$P_{n_1n_2}^{\nu\mu}(z) = z^{n_1-1-\nu}\overline{z}^{n_2-1-\mu} \qquad (4.102)$$

Under an element $g = \begin{pmatrix} a & c \\ b & d \end{pmatrix} P^{\nu\mu}_{n_1 n_2}(z)$ transforms as:

$$P^{\nu\mu}_{n_1 n_2}{}^g(z) = (cz + d)^{n_1 - 1}\overline{(cz + d)}^{n_2 - 1} P^{\nu\mu}_{n_1 n_2}(z) \tag{4.103}$$

It can be shown that these are essentially the only finite-dimensional invariant subspaces of D_χ. For further information see [37] pages 265 and 298.

By constructing the finite-dimensional, invariant subspaces we have now solved one important problem in the theory of representations of SL(2,C). We will now briefly touch the problem if we can introduce a scalar product in the function space under which the representation becomes unitary.

We first recall the definition of a hermitian functional:

Definition 4.12 Let V be a vector space. A map: $\psi : V \times V \to \mathbf{C}$ is called a *hermitian functional* if it is linear in the first component and if $\psi(x, y) = \overline{\psi(y, x)}$.

From [14] (page 190) we get the following result for invariant hermitian functionals on D_χ:

Assume $\chi = (n_1, n_2)$ is an arbitrary index (i.e. $n_i \in \mathbf{C}$ and $n_1 - n_2$ is an integer). Then it can be shown that an invariant hermitian functional exists on a D_χ if and only if $n_1 = -\overline{n_2}$ or $n_1 = \overline{n_2}$.

If the n_i are integers then we find that the condition $n_1 = -\overline{n_2}$ implies $n_1 = 0 = n_2$, this leads thus to the trivial representation space of constant image functions. In the second case we find that $n_1 = n_2 = n$. In this case the functional is given by (see [14] section 6.4):

$$(\varphi, \psi) = (-1)^n \frac{i}{2} \int \varphi^{(n,n)}(z)\overline{\psi}(z)\, dz d\overline{z} \tag{4.104}$$

where

$$\varphi^{(n,n)}(z) = \frac{\partial^{2n}\varphi}{\partial z^n \partial \overline{z}^n}.$$

This functional does not define a scalar product (i.e. a positive, Hermitian functional) on D_χ since D_χ is a space of polynomials of degree less then n in z and $\overline{z}$ and $\varphi^{(n,n)}$ is therefore zero on D_χ.

4.4 Exercises

Exercise 4.1 Define the (2-D) Laplace operator Λ^2 as

$$\Lambda^2 = \frac{\partial^2}{\partial x^2} + \frac{\partial^2}{\partial y^2}$$

Show that Λ^2 is in polar coordinates (r, φ) given by:

$$\Lambda^2 = \frac{\partial^2}{\partial r^2} + \frac{1}{r^2}\frac{\partial^2}{\partial \varphi^2} + \frac{1}{r}\frac{\partial}{\partial r}$$

Exercise 4.2 Define the (3-D) Laplace operator Λ^3 as

$$\Lambda^3 = \frac{\partial^2}{\partial x^2} + \frac{\partial^2}{\partial y^2} + \frac{\partial^2}{\partial z^2}$$

and the spherical Laplace operator Λ as

$$\Lambda = \frac{1}{\sin \theta} \left(\frac{\partial}{\partial \theta} \left(\sin \theta \frac{\partial}{\partial \theta} \right) + \frac{1}{\sin \theta} \frac{\partial^2}{\partial \varphi^2} \right)$$

Show that

$$\Lambda^3 = \frac{1}{r^2 \sin \theta} \left(\frac{\partial}{\partial r} \left(r^2 \sin \theta \frac{\partial}{\partial r} \right) + \frac{\partial}{\partial \theta} \left(\sin \theta \frac{\partial}{\partial \theta} \right) + \frac{\partial}{\partial \varphi} \left(\frac{1}{\sin \theta} \frac{\partial}{\partial \varphi} \right) \right)$$

Exercise 4.3 Prove the statements in theorem 4.1.

Exercise 4.4 Check equation 4.18 for the vector product.

Exercise 4.5 Check the equations 4.20 and 4.23.

Chapter 5

Fourier Series on Compact Groups

In the previous chapters we saw we could identify functions on a homogeneous space (X, G) with a set X and transformation group G with functions on the group G. The most important examples are for us the unit circle together with $SO(2)$ and the unit sphere together with $SO(3)$. In the last chapter (theorem 3.13 and the examples in 3.3) we saw that the irreducible representations of $SO(2)$ are given by the functions $e^{im\phi}$. From the elementary theory of Fourier series it is known that every function on the unit circle can be developed into a Fourier series. Reformulating this in group theoretical language this gives:

"Every function on the homogeneous space X (the unit circle) can be developed into a series of matrix entries of the irreducible representations of $G(= SO(2))$."

In Hilbert space terminology this becomes:

"The matrix entries of the irreducible representations of G form a basis of the L^2 space $L^2(X)$."

In this chapter we will mainly prove that this statement is true for all compact groups. In the second section of this chapter we will then use this result to show some fundamental properties of the characters of a group.

5.1 The Peter-Weyl Theorem

We define the space of all square-integrable functions on a compact group as follows:

Definition 5.1 Let G be a compact group with Haar integral $\int_G f(g)\, dg$. Then we define $L^2(G)$ as the set of complex-valued functions f that satisfy the following conditions:

1. f is measurable under the Haar measure

2. $\int_g |f(g)|^2 \, dg < \infty$

On $L^2(G)$ we define the scalar product:

$$\langle f_1, f_2 \rangle = \int_G f_1(g)\overline{f_2(g)} \, dg$$

With these definitions $L^2(G)$ becomes a Hilbert space.

In this space we define the regular representations of G as:

Definition 5.2   1. The representation $T : G \to L^2(G); h \mapsto T(h)$ with

$$T(h)f(g) = f(gh)$$

is called the *right regular representation* of G.

2. The representation $S : G \to L^2(G); h \mapsto S(h)$ with

$$S(h)f(g) = f(h^{-1}g)$$

is called the *left regular representation* of G.

To see that the definitions are meaningful we have to show that $T(h)f$ is indeed an element of $L^2(G)$ and that T is linear and continuous. We omit the proof since it is purely technical.

Theorem 5.1 The left- and the right-regular representations are unitary.

This can be seen by a simple computation:

$$\langle T(h)f_1, T(h)f_2 \rangle = \int_G f_1(gh)\overline{f_2(gh)} \, dg = \int_G f_1(g)\overline{f_2(g)} \, dg = \langle f_1, f_2 \rangle$$

Theorem 5.2 The left- and the right regular representations of G are unitarily equivalent, i.e. there is a one-to-one mapping $A : L^2(G) \to L^2(G)$ such that A is unitary and $AS(g) = T(g)A$ for all $g \in G$.

Define $Af(g) = f(g^{-1})$ and find for $TA(g)$:

$$
\begin{aligned}
T(g)Af(h) &= T(g)f(h^{-1}) = f(h^{-1}g) = f((g^{-1}h)^{-1}) & (5.1) \\
&= A(f(g^{-1}h)) = A(S(g)f(h)) & (5.2)
\end{aligned}
$$

We now investigate the matrix entries of finite-dimensional, irreducible, unitary representations of a compact group G. In the rest of this section we will assume that the following conditions are satisfied:

We assume that the group G is compact, by T we denote a finite-dimensional, irreducible, unitary representation of G in a Hilbert space H. We fix an orthonormal basis in H and denote the elements of this basis by $e_1, ..., e_n, ...$ In this basis $T(g)$ is described by a unitary matrix and we will also denote this matrix by $T(g)$. The matrix elements of T are denoted by $t_{ij}(g)$ and as functions on G they are continuous. If we want to investigate different representations then we use a superscript to distinguish them. T^α is thus the representation belonging to the index α and $t^\alpha_{ij}(g)$ are the matrix entries of T^α. The dimension of the representation T^α will be denoted by $n_\alpha = \dim T^\alpha$. If A is some index set such that for all elements $\alpha \in A$ we have a finite, irreducible, unitary representation T^α then we say that the set $\{T^\alpha | \alpha \in A\}$ is a *complete set of finite-dimensional representations* if all finite-dimensional, irreducible, unitary representations of G are equivalent to one of the elements in $\{T^\alpha | \alpha \in A\}$. For the matrix entries of finite-dimensional, irreducible, unitary representations we have the following theorem:

Theorem 5.3   1. If T^α and $T^{\alpha'}$ are two finite-dimensional, irreducible, unitary representations then we have the following relations for their matrix entries:

$$\int_G t^\alpha_{ij}(g) t^{\alpha'}_{kl}(g)\, dg = \begin{cases} 0 & \text{if } i \neq k \text{ or } j \neq l \text{ or } \alpha \neq \alpha' \\ 1/n_\alpha & \text{if } i = k \text{ and } j = l \text{ and } \alpha = \alpha' \end{cases} \tag{5.3}$$

The relations in equation 5.3 are called the *orthogonality relations for a compact group*.

2. The functions $e^\alpha_{ij}(g) = \sqrt{n_\alpha} t^\alpha_{ij}(g)$ with $\alpha \in A, 0 \leq i, j \leq n_\alpha$ form an orthonormal system in $L^2(G)$.

Proof: Assume H and H' are the representation spaces of T^α and $T^{\alpha'}$. Denote further the space of all linear mappings from H to H' by L. For an operator $K \in L$ we define the two new operators:

$$K_1(g) = T^\alpha(g) K T^{\alpha'}(g^{-1})$$

and

$$K_2 = \int_G K_1(g)\, dg = \int_G T^\alpha(g) K T^{\alpha'}(g^{-1})\, dg$$

The operators K_1 and K_2 are also elements of L and we have furthermore

$$T^\alpha(h) K_2 = K_2 T^{\alpha'}(g)$$

since

$$
\begin{aligned}
T^\alpha(h) K_2 &= T^\alpha(h) \int_G K_1(g)\, dg & (5.4) \\
&= T^\alpha(h) \int_G T^\alpha(g) K T^{\alpha'}(g^{-1})\, dg \\
&= \int_G T^\alpha(h) T^\alpha(g) K T^{\alpha'}(g^{-1}) T^{\alpha'}(h^{-1})\, dg\, T^{\alpha'}(h) \\
&= \int_G T^\alpha(hg) K T^{\alpha'}((hg)^{-1})\, dg\, T^{\alpha'}(h) \\
&= \int_G T^\alpha(g) K T^{\alpha'}((g)^{-1})\, dg\, T^{\alpha'}(h) \\
&= \int_G K_1(g)\, dg\, T^{\alpha'}(h) = K_2 T^{\alpha'}(h)
\end{aligned}
$$

If T^α and $T^{\alpha'}$ are two irreducible, inequivalent representations then we get from Schur's lemma 3.4 that $K_2 = 0$, i.e.

$$\int_G T^\alpha(g) K T^{\alpha'}(g^{-1})\, dg = 0 \tag{5.5}$$

for all linear maps K. Rewrite the last equation 5.5 with the matrix entries and get:

$$\int_G \sum_{\mu=1}^{n_\alpha} \sum_{\nu=1}^{n_{\alpha'}} t^\alpha_{j\mu}(g) k_{\mu\nu} t^{\alpha'}_{\nu i}(g^{-1})\, dg = 0 \tag{5.6}$$

for all $j = 0, \ldots, n_\alpha$ and all $i = 0, \ldots, n_{\alpha'}$. Using the identity: $t^{\alpha'}_{\nu i}(g^{-1}) = \overline{t^{\alpha'}_{i\nu}(g)}$ equation 5.6 becomes:

$$\int_G \sum_{\mu=1}^{n_\alpha} \sum_{\nu=1}^{n_{\alpha'}} t^\alpha_{j\mu}(g) k_{\mu\nu} \overline{t^{\alpha'}_{i\nu}(g)}\, dg = 0 \tag{5.7}$$

For the function $k_{\mu\nu}$ defined as

$$k_{\mu\nu} = \left\{ \begin{array}{ll} 1 & \text{for } \mu = m \text{ and } \nu = n \\ 0 & \text{for all other indices} \end{array} \right. \tag{5.8}$$

we get $\left\langle t^\alpha_{jm}(g), t^{\alpha'}_{in}(g) \right\rangle = 0$. This shows that the matrix entries are orthogonal if the representations are inequivalent. Now assume that $T^{n_\alpha} = T^{n'_\alpha}$ and write T instead of T^{n_α} and n instead of n_α. In this case we have $T(h)K_2 = K_2 T(g)$ and from Schur's lemma we get $K_2 = \lambda id$. From the property of the trace we find:

$$\begin{aligned} \text{tr } K_2 &= \text{tr } \int_G T(g) K T(g^{-1})\, dg & (5.9) \\ &= \text{tr } \int_G T(g) K (T(g))^{-1}\, dg \\ &= \int_G \text{tr } T(g) K (T(g))^{-1}\, dg \\ &= \int_G \text{tr } K\, dg = \text{tr } K \end{aligned}$$

But since $K_2 = \lambda id$ we find $\text{tr } K_2 = \lambda \cdot n$ and therefore $\lambda = \text{tr } K_2/n$ and $K_2 = \lambda id = \frac{\text{tr } K_2}{n} id$. Selecting the operators K as in the equation 5.8 we find the expressions in 5.3 for the case where $\alpha = \alpha'$.

In the main theorem of this section we show that the matrix entries define not only an orthonormal system in $L^2(G)$ but that this system is also complete:

Theorem 5.4 [Peter-Weyl] Let the functions $e^\alpha_{ij}(g)$ be defined as in theorem 5.3. Then the system $\left\{ e^\alpha_{ij}(g) \right\}_\alpha$ is complete in $L^2(G)$.

Proof: From the theory of Hilbert spaces it is known that we have to show that for every element $f(g) \in L^2(G)$ and every constant $\epsilon > 0$ there is a finite number of matrix entries $t^\alpha_{ij}(g)$ such that

$$\left\| f(g) - \sum_{i,j,\alpha} \gamma^\alpha_{ij} t^\alpha_{ij}(g) \right\| < \epsilon.$$

Take a real-valued continuous function χ, not identically equal to zero, such that $\chi(g^{-1}) = \chi(g)$ (for example $\chi(g) + \chi(g^{-1})$ for any non-negative function χ) and define $K(g_1, g_2) = \chi(g_1 g_2^{-1})$. This is a symmetric continuous function on $G \times G$ and we define the operator k as

$$\phi(g) \mapsto k(\phi)(g) = \psi(g) = \int_G K(g, g_1) \phi(g_1)\, dg_1 \tag{5.10}$$

The kernel K is quadratically integrable:

$$\int_{G \times G} |K(g, h)|^2\, dg\, dh < \infty \tag{5.11}$$

and from the Hilbert-Schmidt theory of integral equations we find that every function ψ of the form $\psi = k(\phi)$ is the sum of an absolutely and uniformly convergent series of eigenfunctions of the eigenvalue problem

$$k(\phi)(g) = \int_G K(g, g_1) \phi(g_1)\, dg_1 = \lambda \phi(g). \tag{5.12}$$

From the Hilbert-Schmidt theory it is also known that k has it least one eigenvalue λ and that the space $M_\lambda \subset L^2(G)$ consisting of the zero function and all eigenfunctions belonging to the eigenvalue λ forms a subspace of $L^2(G)$. This subspace is invariant under right translations. This can be easily seen by inserting $\phi(gg_0)$ into the equation 5.12 and computing the integral using the definition of K and the right invariance of the Haar integral.

M_λ is thus the representation space of a subrepresentation T_λ of the right regular representation. T_λ is continuous, finite-dimensional and unitary and can therefore be broken up into a finite number of irreducible unitary representations $T_\lambda^{(1)}, \ldots, T_\lambda^{(p)}$ and M_λ is the direct sum of the representation spaces $M_\lambda^{(1)}, \ldots, M_\lambda^{(p)}$).

We show now that all functions in $M_\lambda^{(k)}$ are linear combinations of the matrix entries of the representation $T_\lambda^{(k)}$:

Select a set of basis functions $e_i(g), i = 1, \ldots n$ in $M_\lambda^{(k)}$ and denote the matrix entries of the representation $T_\lambda^{(k)}$ by $c_{ij}(g)$. Since $T_\lambda^{(k)}$ is a subrepresentation of the right regular representation we have $T_\lambda^{(k)}(g_0)f(g) = f(g_0g)$ for all $f \in M_\lambda^{(k)}$ and especially for the basis elements $e_i(g)$. Taking the identity $e \in G$ as g_0 we find that for every element $g \in G$ we have

$$e_i(g) = e_i(eg) = \sum_{k=0}^{n} c_{ki}(g)e_k(e). \tag{5.13}$$

The $e_k(e)$ are constants and the basis elements $e_i(g)$ are therefore linear combinations of the matrix entries c_{ij}. All elements in $M_\lambda^{(k)}$ are thus linear combinations of the matrix entries and we have thus shown that every element $\psi \in L^2(G)$ of the form 5.10 is a linear combination of the $c_{ij}(g)$.

We show now that every continuous function on G can be uniformly approximated by functions of the form 5.10.

We only sketch the proof: For a given continuous function f and a constant $\epsilon > 0$ we select a neighborhood U of $e \in G$ such that $|f(g) - f(g')| < \epsilon$ for all $g, g' \in G$ such that $g' \in gU$. We can also assume that $U = U^{-1}$. Then we select a neighborhood V of e such that the closure of V is contained in U. Now choose a nonnegative function ψ such that $\psi = 1$ on V and $\psi = 0$ outside U. From ψ we construct $\chi = c(\psi(g) + \psi(g^{-1}))$ such that $\int_G \chi(g)\, dg = 1$.

Now we define $\phi(g)$ as

$$\phi(g) = \int_G f(g_1)\chi(gg_1^{-1})\, dg_1$$

Setting $h = g_1g^{-1}$ and using the invariance of the integral gives

$$\phi(g) = \int_G f(hg)\chi(h^{-1})\, dh$$

and from

$$1 = \int_G \chi(g)\, dg = \int_G \chi(h^{-1})\, dh$$

we get

$$f(g) = \int_G f(g)\chi(h^{-1})\, dh.$$

$\phi(g)$ is a function of the type defined in 5.10 and we will show that $|f(g) - \phi(g)| < \epsilon$:

$$
\begin{aligned}
|f(g) - \phi(g)| &= \left| \int_G f(g)\chi(h^{-1})\, dh - \int_G f(hg)\chi(h^{-1})\, dh \right| \qquad (5.14) \\
&= \left| \int_G (f(g) - f(hg))\,\chi(h^{-1})\, dh \right| \\
&\leq \int_G |(f(g) - f(hg))|\,\chi(h^{-1})\, dh \\
&= \int_U |(f(g) - f(hg))|\,\chi(h^{-1})\, dh \\
&< \int_U \epsilon\chi(h^{-1})\, dh = \epsilon
\end{aligned}
$$

To complete the proof we notice that the set of continuous functions on G is dense in $L^2(G)$, i.e. for every $\epsilon > 0$ and every $f \in L^2(G)$ we can find a continuous function $\tilde{f}$ such that $\left\| f - \tilde{f} \right\| < \epsilon/2$. For $\tilde{f}$ we can find a linear combination ϕ of matrix entries such that $\left\| \tilde{f} - \phi \right\| < \epsilon/2$ which gives the desired approximation of f by a finite sum of matrix entries.

As a simple consequence of this theorem we get the following expansion of functions in $L^2(G)$:

Theorem 5.5   1. Every function $f \in L^2(G)$ can be written in the form:

$$
\begin{aligned}
f &= \sum_{\alpha \in A} \sum_{l,j} \left\langle f, e_{lj}^{\alpha} \right\rangle e_{lj}^{\alpha} \qquad (5.15) \\
&= \sum_{\alpha \in A} \sum_{l,j} n_{\alpha} \left\langle f, t_{lj}^{\alpha} \right\rangle t_{lj}^{\alpha} \qquad (5.16)
\end{aligned}
$$

2. For every function $f \in L^2(G)$ we have the *Plancerel formula:*

$$
\begin{aligned}
\langle f, f \rangle &= \sum_{\alpha \in A} \sum_{l,j} \left| \left\langle f, e_{lj}^{\alpha} \right\rangle \right|^2 \qquad (5.17) \\
&= \sum_{\alpha \in A} \sum_{l,j} n_{\alpha} \left| \left\langle f, t_{lj}^{\alpha} \right\rangle \right|^2 \qquad (5.18)
\end{aligned}
$$

We will only mention that in a sense also the converse of the Peter-Weyl theorem is true (for a proof see [37].)

Theorem 5.6 Let $\{T^{\alpha}, \alpha \in A\}$ be a set of irreducible, unitary representations of a compact group G. Let $\left\{ t_{ij}^{\alpha}, \alpha \in A, i, j = 1, ..., \dim T^{\alpha} \right\}$ be a set of matrix elements of the representations T^{α} in some orthonormal basis. If the finite linear combinations of these matrix elements form a dense subset of the space $L^2(G)$ (or of the space $C(G)$ of continuous functions on G), then every irreducible, unitary representation of G is unitarily equivalent to one of the representations in $\left\{ t_{ij}^{\alpha}, \alpha \in A, i, j = 1, ..., \dim T^{\alpha} \right\}$.

5.2 Characters

The Peter-Weyl theorem and the characters of a group can be used to find a characterization of the irreducible representations of a compact group. We recall first the definition of a group character:

Let T be a finite-dimensional representation of a compact group G and let $e_1, ... e_n$ be a basis of the representation space H. The matrix elements of T in this basis will be denoted by $t_{ij}(g)$. The character of the representation T is defined as the trace of the representation matrices:

$$\chi(g) = \sum_{i=1}^{n} t_{ii}(g) = \text{tr } (t_{ij}(g)) \qquad (5.19)$$

The character of a representation depends only on the representation and not on the selected basis e_i, since the trace of a matrix is independent of the basis of the underlying space. From the properties of the trace it follows also that the character does not change if we replace T by an equivalent representation and that the character is constant on conjugacy classes in G. For the characters we find the following *orthogonality relations for characters:*

Theorem 5.7 The characters χ^α and $\chi^{\alpha'}$ of irreducible unitary representations T^α and $T^{\alpha'}$ of the compact group G satisfy the orthogonality relations:

$$\left\langle \chi^\alpha, \chi^{\alpha'} \right\rangle = \int_G \chi^\alpha(g) \overline{\chi^{\alpha'}(g)} \, dg = \begin{cases} 0 & \text{if } T^\alpha \text{ is not equivalent } T^{\alpha'} \\ 1 & \text{if } T^\alpha \text{ is equivalent } T^{\alpha'} \end{cases}$$

This can be easily seen by inserting the definition of a character in the integral and using the orthogonality of the matrix elements.

For the character of a finite-dimensional unitary representation we get the following characterization:

Theorem 5.8 Let S be a finite-dimensional, unitary representation and assume that S is the direct sum of the irreducible, unitary representations $T_k^\alpha, k = 1, ..., p$:

$$S = n_1 T^{\alpha_1} + ... + n_p T^{\alpha_p}. \qquad (5.20)$$

with natural numbers n_k. Denote the character of S by χ and the character of T_k^α by χ_k. Then we have

$$\chi = n_1 \chi_1 + ... + n_p \chi_p \qquad (5.21)$$

with

$$n_k = \langle \chi, \chi_p \rangle \qquad (5.22)$$

Conversely if the character χ of a representation allows a decomposition of the type shown in equation 5.21 then S has a decomposition of the form 5.20.

Theorem 5.9 The characters of two finite-dimensional unitary representations of a compact group G coincide if and only if the representations are equivalent.

Proof: Denote the characters by T^1 and T^2 and the characters by χ^1 and χ^2. If T^1 and T^2 are equivalent then are the characters identical. Conversely if $\chi^1 = \chi^2$ then the decompositions of type 5.21 coincide and so we find that also the representations coincide.

Theorem 5.10 A continuous, finite-dimensional, unitary representation of a compact group G is irreducible if and only if $\int_G |\chi(g)|^2 \, dg = 1$.

Proof: If χ is irreducible then we get $\int_G |\chi(g)|^2 \, dg = 1$ from theorem 5.3. Now assume $\int_G |\chi(g)|^2 \, dg = 1$ then we find also from theorem 5.3

$$1 = \langle \chi, \chi \rangle = n_1^2 \langle \chi_1, \chi_1 \rangle + \ldots + n_k^2 \langle \chi_k, \chi_k \rangle = n_1^2 + \ldots + n_k^2$$

for natural numbers n_k and characters of irreducible representations.

We will now apply these results to show that the system of irreducible representations of $SO(3)$ that was constructed in the last chapter is indeed complete.

In the beginning of this section we mentioned already that the character of a representation is constant on conjugacy classes. In the case of $SO(3)$ this means that $\chi(g)$ is only a function of the rotation angle of g since we can consider $h^{-1}gh$ instead of g and select h such that h exchanges the rotation axis of g with the z-axis. It can be easily seen that g and $h^{-1}gh$ have the same rotation angle and that $h^{-1}gh$ is a rotation around the z-axis. We conclude that it is sufficient to compute the character value only for rotations around the z-axis. If we consider the representation in the space of surface harmonics $Y_l^{-l}, \ldots, Y_l^l$ then we find from the form of the surface harmonics that the character χ_l of this representation has the form:

$$\chi_l(g) = \chi_l(\alpha) = \sum_{m=-l}^{l} e^{im\alpha} = \frac{\sin(l + 1/2)\alpha}{\sin \alpha/2} \tag{5.23}$$

where α is the rotation angle of the rotation g (the last equation is obtained by using the formula for an geometric sum). For the product of two characters χ_l and χ_m we find

$$\chi_l(g)\chi_m(g) = \frac{\cos(l - m)\alpha - \cos(l + m + 1)\alpha}{1 - \cos \alpha} \tag{5.24}$$

The functions χ_l are orthonormal on the group $SO(3)$ and we can therefore rewrite the Haar integral for the characters as

$$\int_{SO(3)} \chi_l(g)\chi_m(g) \, dg = \int_0^\pi \chi_l(\alpha)\chi_m(\alpha)w(\alpha) \, d\alpha \tag{5.25}$$

for some weight function $w(\alpha)$. From the orthonormality of the characters one gets:

$$w(\alpha) = \frac{1 - \cos \alpha}{\pi} \tag{5.26}$$

and the Haar integral in equation 5.25 becomes thus:

$$\int_{SO(3)} \chi_l(g)\chi_m(g) \, dg = \int_0^\pi \chi_l(\alpha)\chi_m(\alpha)\frac{1 - \cos \alpha}{\pi} \, d\alpha \tag{5.27}$$

We use equation 5.27 now to show that the representations in the spaces $Y_l^{-l}, \ldots, Y_l^l$ are indeed a complete set of finite-dimensional, irreducible, unitary representations of $SO(3)$.

If χ is the character of any finite-dimensional irreducible unitary representation of $SO(3)$ then we have $\langle \chi, \chi \rangle = 1$. If χ is different from all χ_l then we have also $\langle \chi, \chi_l \rangle = 0$ for all l. To see that the surface harmonics form indeed a complete set of irreducible representations we have to show that this is impossible:

We use $1 - \cos \alpha = 2(\sin \frac{1}{2}\alpha)^2$ and get:

$$\langle \chi, \chi_l \rangle = \int_0^\pi \chi(\alpha)\chi_l(\alpha)\frac{1 - \cos \alpha}{\pi} \, d\alpha \tag{5.28}$$
$$= \frac{2}{\pi} \int_0^\pi \chi(\alpha)\frac{\sin(l + 1/2)\alpha}{\sin \alpha/2}(\sin \alpha/2)^2 \, d\alpha$$
$$= \frac{2}{\pi} \int_0^\pi \chi(\alpha) \sin((l + 1/2)\alpha) \sin(\alpha/2) \, d\alpha$$

Making the substitution $t = \alpha/2$ and $F(t) = \sin \frac{\alpha}{2}\chi(\alpha)$ we find

$$\langle \chi, \chi_l \rangle = \frac{4}{\pi} \int_0^{\pi/2} F(t) \sin(2l + 1)t \, dt \tag{5.29}$$

Next define $F^-(t)$ as:

$$F^-(t) = \begin{cases} F(t) & \text{if } 0 \leq t \leq \pi/2 \\ F(t - \pi/2) & \text{if } \pi/2 \leq t \leq \pi/2 \\ -F^-(t) & \text{if } 0 \leq -t \leq \pi. \end{cases} \tag{5.30}$$

Then F^- is odd around the origin and even around the points $\pi/2$ and $-\pi/2$. Furthermore F^- is orthogonal to all functions $\sin(2l + 1)t$ and we find therefore that F^- is identically zero. F is thus identically zero and thus also χ. This shows that the representations in the space of the surface harmonics exhaust all the finite-dimensional representations of $SO(3)$. Using the fact that all irreducible unitary representations of a compact group are finite-dimensional (see 3.12) we see that we found all irreducible unitary representations of $SO(3)$.

The complex exponentials and the surface harmonics form thus complete sets of basic functions for the functions on the unit circle and the unit sphere respectively:

Theorem 5.11 1. The functions $e^{in\phi}$ form a complete orthonormal system for the square-integrable functions on the unit circle.

2. The surface harmonics Y_l^m form a complete orthonormal system for the square-integrable functions on the unit sphere.

The proof is very simple: Assume f is a function on the circle (sphere) that is orthonormal to all complex exponentials (surface harmonics). Then we consider the spaces $\{f(g(x))|g \in SO(2)\}$ and $\{f(g(x))|g \in SO(3)\}$ respectively. These spaces are invariant under the rotation groups and orthogonal to all exponentials (surface harmonics). These spaces define thus a representation of the group and they contain thus finite-dimensional subspaces that are invariant under the group operation. Since these subspaces are orthogonal to all exponentials (surface harmonics) we see that we have found a new, irreducible, unitary finite-dimensional representation of the rotation group. This is impossible and therefore we have $f = 0$.

Chapter 6

Applications

In the last chapters we developed the basic facts about representations of groups. In some special cases (see theorem 3.13, examples 3.3 and chapter 4) we derived also a description of all irreducible representations of a given group. In this chapter we describe how we can use these results to investigate some problems from image science. After investigating the general theory we will illustrate it with some examples from image reconstruction, filter design and neural networks.

6.1 Eigenvectors of Intertwining Operators

In this section we describe the eigenvalue problem in its most general formulation. In the following we will always assume that H is a Hilbert space. By $GL(H)$ we denote as usual the set of invertible linear operators that map H to H. In the following theorem we describe the connection between the eigenvectors of an operator and the representations of a group.

Theorem 6.1 Assume that $T : G \to GL(H)$ is a representation of G and that the function $A : H \to H$ is an intertwining operator (i.e. $AT(g) = T(g)A$ for all $g \in G$). We denote by $H_\lambda = \{x \in H : Ax = \lambda x\}$ the space of eigenvectors of A belonging to the eigenvalue λ. For every $g \in G$ we define $T_\lambda(g)$ as the restriction of $T(g)$ to H_λ. Under these conditions is $T_\lambda : G \to GL(H_\lambda)$ a representation of G.

To see that T_λ really is a representation of G we only have to show that $T_\lambda(g) \in GL(H_\lambda)$ for all g. The other properties of a representation are all inherited from the original representation T. We show now that $T(g)$ does indeed map H_λ to H_λ :

Using the intertwining property we find for an element $f \in H_\lambda$:

$$A(T(g)f) = (AT(g))f = (T(g)A)f = T(g)(Af) = T(g)(\lambda f) = \lambda T(g)f$$

The image $T(g)f \in H_\lambda$ of an eigenvector $f \in H_\lambda$ is thus also an eigenvector in H_λ.

We can use the information that H_λ is a representation space of G to investigate its structure. For example if the space H_λ is finite-dimensional then we know that representations on Hilbert spaces are completely reducible and H_λ can thus be decomposed into a sum of subspaces where each subspace defines an irreducible representation of G. In the previous chapter we saw that for some groups it is possible to compute all their irreducible representations. Using these two results we can thus from the intertwining property draw some conclusions about the form of the eigenvectors. How this works in practice is demonstrated in the remaining sections of this chapter.

In our applications we will normally not consider arbitrary Hilbert spaces H but we will mainly deal with Hilbert spaces of square-integrable functions on sets that are essentially homogeneous spaces of the given group G.

More precisely we assume that the Hilbert space H is given by the L^2-space $L^2(\mathcal{D})$ where $\mathcal{D}$ is a topological space of the form $\mathcal{D} = R \times X$ with X a closed homogeneous space of G. R is a set and in our applications it is normally an interval or the real axis. The most important examples are the unit disk and the unit ball. The groups G are in these cases the rotation groups $SO(2)$ and $SO(3)$, the homogeneous spaces are the unit circle and the unit sphere and R is the unit interval $[0, 1]$. Decomposing $\mathcal{D}$ into a product of type $R \times X$ is similar to the introduction of polar coordinates consisting of an angular part (the space X) and a radial part (the set R).

If the group G operates on an arbitrary set $\mathcal{D}$ then it is however in general not possible to decompose $\mathcal{D}$ into a product of the form $R \times X$ as the following example shows (see [4]): We take $\mathcal{D}$ as the unit square. The group G consists of the transformations $g_t(x, y) = (\xi, \eta)$ with $\xi = x + ta \bmod 1$ and $\eta = y + tb \bmod 1$ where a and b are two fixed numbers such that a/b is not a rational number. Then it can be shown that the orbit of a fixed point (x_0, y_0) under the group $\{g_t(x_0, y_0) : t \in \mathbf{R}\}$ is dense in $\mathcal{D}$. The closure of the orbit is thus equal to the whole unit square. The orbit is however not equal to $\mathcal{D}$.

If $\mathcal{D}$ is of the form $\mathcal{D} = R \times X$ then we construct a representation of G as follows: From the introduction of homogeneous spaces in section 2.2.5 we know that X can be identified with a factor group G/G_0 where G_0 is a subgroup of G. If g is an element in G then we denote the coset $G_0 g \in G/G_0$ by $\tilde{g}$. A point in the domain $\mathcal{D}$ can thus be written as a pair $(r, \tilde{g})$ where r is an element in R and $\tilde{g}$ is an element in G/G_0. The functions f in $H = L^2(\mathcal{D})$ have thus the form $f(r, \tilde{g})$. The representations used will mainly be the left or the right regular representations (see 5.2)). For example if h is an element $h \in G$ then we may define the operator: $T(h)f(r, \tilde{g}) = f(r, \widetilde{gh})$.

In our standard example we use $G = SO(2)$. The unit circle is a homogeneous space of $SO(2)$ and as our domain $\mathcal{D}$ we select the unit circle. A point in the domain can be described with an angular variable φ and if ψ is the rotation angle of a certain rotation g then the transformation of the domain is just a translation $\varphi \mapsto \varphi - \psi$. The linear map $T(\psi)$ maps a function $f(\varphi)$ to the rotated function $f(\varphi - \psi)$.

An irreducible representation defines an one-dimensional invariant subspace in H such that $f(\varphi - \psi) = e^{in\psi} f(\varphi)$ for all elements f in this subspace. This is the functional equation of the exponential function and the one-dimensional invariant subspaces consist of the multiples of the exponential functions.

In the case where the domain is the unit disk we find that $\mathcal{D}$ is given by the product $[0, 1] \times SO(2)$, the transformations of the domain are of the form $(r, \varphi) \mapsto (r, \varphi - \psi)$ and the linear map $T(\psi)$ becomes: $f(r, \varphi) \mapsto T(\psi)f(r, \varphi) = f(r, \varphi - \psi)$. By treating the variable $r \in R$ as a parameter we can mainly restrict our attention to functions defined on homogeneous spaces.

6.2 Laplacian and Integral Transforms

In theorem 3.13 we described all irreducible representations of the 2-D rotation group $SO(2)$ and we will illustrate the results from the previous section with operators that intertwine with $SO(2)$. We first investigate the Laplacian Δ that is defined in cartesian coordinates as:

$$\Delta f(x,y) = \frac{\partial^2 f(x,y)}{\partial x^2} + \frac{\partial^2 f(x,y)}{\partial y^2} \tag{6.1}$$

In polar coordinates (r,φ) this becomes:

$$\Delta f(r,\varphi) = \frac{\partial^2 f(r,\varphi)}{\partial r^2} + r^{-2}\frac{\partial^2 f(r,\varphi)}{\partial \varphi^2} + r^{-1}\frac{\partial f(r,\varphi)}{\partial r}. \tag{6.2}$$

Defining the transformed function $f^\psi(r,\varphi)$ as $f(r,\varphi-\psi)$ we see immediately that $\Delta(f^\psi) = (\Delta f)^\psi$, i.e. the Laplacian is really an intertwining operator for $SO(2)$.

The representations of $SO(2)$ are given by the complex exponentials and we find indeed that $\Delta e^{in\varphi} = -n^2 e^{in\varphi}$. In the general case we separate the variables r and φ and expand an arbitrary function f into a series

$$f(r,\varphi) = \sum_n h_n(r)e^{in\varphi}.$$

Inserting this into the expression for the Laplacian we find:

$$\Delta f = \sum_n \left(h_n''(r)e^{in\varphi} - \frac{n^2}{r}h_n(r)e^{in\varphi} + r^{-1}h_n'(r)e^{in\varphi} \right) =$$

$$\sum_n e^{in\varphi}\left(h_n''(r) - \frac{n^2}{r}h_n(r) + r^{-1}h_n'(r) \right)$$

Comparing the coefficients for λf and Δf shows that we can find the eigenfunctions of Δ by solving the equations:

$$h_n''(r) - \frac{n^2}{r}h_n(r) + r^{-1}h_n'(r) = \lambda h_n(r)$$

We saw that we could reduce a two-dimensional problem to a one-dimensional problem by using the symmetry properties of the Laplace operator. This is a characteristic feature for many applications of group theoretical methods.

Next we consider the 2-D Fourier transform $\mathcal{F}$ (see also [44]). It is defined in cartesian coordinates as:

$$(\mathcal{F}f)(u,v) = F(u,v) = \int_{\mathbf{R}^2} f(x,y)e^{-i(xu+yv)}\,dx dy \tag{6.3}$$

and in polar coordinates this becomes:

$$(\mathcal{F}f)(\rho,\psi) = \int_0^\infty \int_0^{2\pi} f(r,\varphi)e^{-i\rho r\cos(\varphi-\psi)}r\,d\varphi dr \tag{6.4}$$

Using one of the equations 6.3 or 6.4 it is easy to see that $\mathcal{F}$ intertwines with $SO(2)$. As in the investigation of the Laplacian we compute the eigenfunctions of $\mathcal{F}$ by expanding a function f into a series of exponentials. We get:

$$(\mathcal{F}f)(\rho,\psi) = \sum_n \int_0^\infty h_n(r)r \int_0^{2\pi} e^{in\varphi}e^{-i\rho r\cos(\varphi-\psi)}\,d\varphi dr \tag{6.5}$$

Using the identity (see [38]):

$$J_n(x) = \frac{1}{2\pi} \int_{-\pi}^{\pi} e^{i(n\phi - x \sin \phi)} \, d\phi \tag{6.6}$$

equation 6.5 becomes:

$$(\mathcal{F}f)(\rho, \psi) = 2\pi \sum_n \int_0^\infty r h_n(r) J_n(r\rho) \, dr \, e^{in\psi} = 2\pi \sum_n (K_n h_n)(\rho) e^{in\psi} \tag{6.7}$$

Inserting this into the equation $\mathcal{F}f = \lambda f$ and comparing termwise we find that the coefficient function $h_n(r)$ must satisfy the equation:

$$\lambda h_n(\rho) = (K_n h_n)(\rho) = \int_0^\infty r h_n(r) J_n(r\rho) \, dr \tag{6.8}$$

Again we have reduced a two-dimensional problem to a one-dimensional one.

The Fourier transformation is only one example of a whole class of problems all connected to integral equations of the type

$$\int_{\mathcal{D}} k(X, Y) f(X) \, d\mu(X) = \lambda f(Y) \tag{6.9}$$

We call k the *kernel* of the integral equation. In the Fourier transform case the variables X and Y are vectors in the n-dimensional space $\mathbf{R}^n$ and the kernel is given by $k(X, Y) = e^{i\langle X, Y \rangle}$ and $d\mu(X) = dX$. In the following we make the following assumptions: We define the space $L^2(\mathcal{D})$ as the space of square-integrable functions defined on the domain $\mathcal{D}$ with respect to the integration given by $d\mu(X)$. On this space we define the operator K as

$$(Kf)(Y) = \int_{\mathcal{D}} k(X, Y) f(X) \, d\mu(X) \tag{6.10}$$

We assume further that there is a group G of transformations $g : \mathcal{D} \to \mathcal{D}$ and the group representation T defined as $T(g)f(X) = f^g(X) = f(g^{-1}(X))$. Finally we assume that the integral and the kernel are invariant under the group operation, i.e.

$$\int_{\mathcal{D}} f(X) \, d\mu(X) = \int_{\mathcal{D}} f(g^{-1}(X)) \, d\mu(X) \tag{6.11}$$

and

$$k(g(X), g(Y)) = k(X, Y) \tag{6.12}$$

for all $g \in G$ and all $X, Y \in \mathcal{D}$. Under these conditions we find that K intertwines with G:

$$
\begin{aligned}
T(g)(Kf(Y)) &= (Kf)(g^{-1}(Y)) = \int_{\mathcal{D}} k(X, g^{-1}(Y)) f(X) \, d\mu(X) \\
&= \int_{\mathcal{D}} k(g(X), Y) f(X) \, d\mu(X) \\
&= \int_{\mathcal{D}} k(X, Y) f(g^{-1}(X)) \, d\mu(g^{-1}(X)) \\
&= \int_{\mathcal{D}} k(X, Y) f(g^{-1}(X)) \, d\mu(X) \\
&= K(T(g)f)(Y) \tag{6.13}
\end{aligned}
$$

and we find that K is an intertwining operator. The solutions of the integral equation 6.9 define thus a representation of G. As in the general case we can use group theoretical results to characterize the eigenspaces of K. Solving an integral equation of the type given in equation 6.10 is thus reduced to checking the conditions 6.11 and 6.12 for the integral and the kernel.

We investigated the eigenfunctions of the Laplacian and the Fourier transform only in the two-dimensional case but it should be clear that similar results can be derived for the three-dimensional Laplacian and Fourier transform. In this case the eigenfunctions are of the form $h_n^m(r)Y_n^m(\theta, \phi)$.

6.3 Scale Space Filtering

In the last section we investigated eigenvalue problems involving the Laplacian operator. In this section we will briefly describe how these results might be used in image processing.

Analysis of images at different spatial scales is known to be an important tool in the processing of images [34], [54], [3]. One of the most popular procedures in this branch of image processing is based on the zero-crossings of the Laplacian. This approach can be summarized as follows:

Let ∇^2 denote the Laplacian

$$\nabla^2 f(x_1, ..., x_n) = \frac{\partial^2 f}{\partial x_1^2} + ... + \frac{\partial^2 f}{\partial x_n^2}, \tag{6.14}$$

let $F_\sigma(y) = F(\sigma, y)$ be the Gaussian filter function

$$F(\sigma, y) = \frac{1}{\sigma^n} exp(\frac{-||y||^2}{2\sigma^2}) \tag{6.15}$$

and let $I(y)$ be the image. Then the image filtered by the Laplacian of the Gaussian is given by

$$E(x, \sigma) = \nabla^2 [F(\sigma, y) \star I(y)](x). \tag{6.16}$$

The convolution is denoted by $\star$ as usual and x and y are n-dimensional vectors. In this context σ is variable which controls the amount of blurring. The points x_0 which satisfy

$$E(x_0, \sigma) = 0 \tag{6.17}$$

are the border points at the scaling level defined by σ. Such a point is called a zero-crossing of the Laplacian.

The basic idea is that a large value of σ will smooth out the minor details and a following analysis will therefore only be carried out on the main structures in the image. After this step the analysis is refined by analysing the image with a smaller value of σ.

One basic requirement when designing the filter function F in equation 6.16 is the condition that no new zero-crossings must be created if the amount of blurring controlled by σ is increased. In [54], [3] it was shown that this requirement is fulfilled if $E(x, \sigma)$ is a solution of the n-dimensional heat equation

$$E_\sigma = \frac{\partial E}{\partial \sigma} = a\nabla^2 E \tag{6.18}$$

for a constant a.

6.3.1 The Heat Equation

In the next few equations we will summarize some facts about solutions of the heat equation in two- and three-dimensional space. In contrast to our previous discussion of the Laplacian operator we will now additionally require that the solutions also satisfy certain boundary conditions.

From mathematical physics ([49]) we cite the following result:

Theorem 6.2 Denote by $E|_U$ the restriction of E on the border of the unit circle (2-D) or the surface of the unit sphere (3-D). Then we find that the formal solutions of the heat equation under the boundary conditions

$$E_\sigma = a^2 \nabla^2 E; \quad E|_U = 0; \quad E(x,0) = f \tag{6.19}$$

are given by

$$E(r,\varphi,\sigma) = \frac{1}{\pi} \sum_{k=0}^{\infty} \sum_{j=1}^{\infty} a_{kj} exp\left(-\left[\mu_j^{(k)}\right]^2 a^2\sigma\right) \frac{J_k(\mu_j^{(k)}r)}{\left[J_k'(\mu_j^{(k)})\right]^2} exp(ik\varphi) \tag{6.20}$$

and

$$E(r,U,\sigma) = \frac{1}{\sqrt{r\pi}} \sum_{l=0}^{\infty} \sum_{j=1}^{\infty} \sum_{m=-l}^{l} a_{ljm} exp\left(-\left[\mu_j^{(l+1/2)}\right]^2 a^2\sigma\right) \times$$
$$\times \left(\frac{2l+1}{1+\delta_{0m}}\right)\left(\frac{(l-|m|)!}{(l+|m|)!}\right) \frac{J_{l+1/2}(\mu_j^{(l+1/2)}r)}{\left[J_{l+1/2}'(\mu_j^{(l+1/2)})\right]^2} S_n^m(U)$$

in the 2-D and 3-D case respectively. J_ν is, as usual, the ν-th Bessel function. The coefficients are computed from the start function f by

$$a_{kj} = \int_0^1 \int_0^{2\pi} f(r,\varphi) J_k(\mu_j^{(k)}r) exp(-ik\varphi) r \, dr d\varphi \tag{6.21}$$

and

$$a_{ljm} = \int_0^1 \int_0^\pi \int_0^{2\pi} f(r,\varphi,\theta) J_{l+1/2}(\mu_j^{(l+1/2)}r) Y_l^m(\varphi,\theta) r^{3/2} \sin(\theta) \, dr d\theta d\varphi \tag{6.22}$$

Here (r,θ,φ) are the coordinates of (r,U) in the usual 3-D polar coordinate system and $S_n^m(U) = Y_n^m(\theta,\varphi)$. Furthermore $\mu_j^{(k)}$ is the j-th positive zero of the Bessel function J_k. These series expansions have the advantage that they separate the influence of the scaling parameter σ, the radius r and the angular variables.

For our interpretation of these results it is important to know that these zeros are interlaced (see [50]) which means that if $\nu \geq 0$ then

$$0 < \mu_1^{(\nu)} < \mu_1^{(\nu+1)} < \mu_2^{(\nu)} < \mu_2^{(\nu+1)} < \mu_3^{(\nu)} < \dots \tag{6.23}$$

If we now look at the behavior of E as a function of σ then we see that large values of σ suppress the influence of a_{kl} and a_{ljm} according to the rule given by equations 6.23 and 6.20, 6.21.

6.3.2   Relation between Filtered Image and Original

We have seen that the functions

$$X_{kj}(r,\varphi) = c_{kj} J_k \left(\mu_j^{(k)} r\right) exp(ik\varphi) \qquad \text{(2-D)} \qquad (6.24)$$

and

$$X_{ljm}(r,\theta,\varphi) = \frac{c_{ljm}}{\sqrt{r}} J_{l+1/2} \left(\mu_j^{(l+1/2)} r\right) Y_l^m(\theta,\varphi) \qquad \text{(3-D)} \qquad (6.25)$$

play an important role in studying the function $E = \nabla^2(F \star I)$ (for the definition of the constants c_{kj}, c_{ljm} see [49]). In this section we will now investigate what can be said about the image function I from the knowledge of E, i.e. we will describe in detail the connection between I and E. Here we can again think of F as the Gaussian filter but our discussion is not limited to this case. For ease of notation we will use the symbol X_K to denote both the 2-D case $X_K = X_{kj}$ and the 3-D case $X_K = X_{ljm}$.

The important property of the functions X_K is the fact that they are the eigenfunctions of the Laplacian. More exactly we have the following theorem (see [49]):

Theorem 6.3 The functions X_K are solutions of the boundary problems

$$\nabla^2 f = -\lambda f \qquad f|_U = 0 \qquad (6.26)$$

where, as before, $f|_U$ denotes the restriction of f to the boundary of the unit circle and the unit sphere respectively. The eigenvalues are given by

$$\lambda_{kj} = \left[\mu_j^{(k)}\right]^2 \text{ in 2-D and } \lambda_{lj} = \left[\mu_j^{(l+1/2)}\right]^2 \text{ in 3-D} \qquad (6.27)$$

On the other hand we know from the theory of Bessel functions (see [50]) that functions g satisfying $\nabla^2 g = -k^2 g$ are easily integrable. We have (in 3-D) the following theorem:

Theorem 6.4 Let F be a continuous function of r and let g be a solution of $\nabla^2 g = -k^2 g$ then (under fairly general conditions on F and g) we have

$$\int_{-\infty}^{\infty} \int_{-\infty}^{\infty} \int_{-\infty}^{\infty} g(\xi,\eta,\zeta) F \left(\sqrt{(\xi-x)^2 + (\eta-y)^2 + (\zeta-z)^2}\right) d\xi d\eta d\zeta =$$
$$\frac{4\pi g(x,y,z)}{k} \int_0^{\infty} F(r) r \sin kr \, dr = c_k(F) g(x,y,z)$$

In the 2-D case we find a similar expression by using the property that if $\nabla^2 g = -k^2 g$ then

$$\int_{-\pi}^{\pi} g(r,\varphi) \, d\varphi = 2\pi g(0,0) J_0(kr) \qquad (6.28)$$

which gives:

$$\int_{-\infty}^{\infty} \int_{-\infty}^{\infty} g(\xi,\eta) F \left(\sqrt{(\xi-x)^2 + (\eta-y)^2}\right) d\xi d\eta =$$
$$2\pi g(x,y) \int_0^{\infty} F(r) J_0(kr) r \, dr = c_k(F) g(x,y)$$

This last theorem states that the eigenfunctions of the Laplacian are also eigenfunctions of the convolution with the function F, or, equivalently, that the eigenfunctions of the Laplacian are eigenfunctions of the operator with the kernel function F.

The next thing we need to know is the fact that the complex exponentials and the surface harmonics form a complete orthogonal system for the functions defined on the unit circle and the surface of the unit sphere respectively. The Bessel functions also form an orthogonal system for functions defined on the unit interval:

Theorem 6.5 If $(f,g) = \int_0^1 f(r)g(r)r\ dr$ then the functions $\left\{ J_\nu \left(\mu_k^{(\nu)} r \right) \right\}_{k=1,\ldots}$ form an orthogonal system of functions defined in the unit interval [0 1] for a fixed value of ν.

Now we are ready to establish the desired connection between the original intensity distribution I and its filtered versions:

We start by developing the function I defined inside the unit sphere (circle) into a series. Theorem 6.5 and the Peter-Weyl theorem state that this is always possible.

$$I(\xi) = \sum_K b_K X_K(\xi) \tag{6.29}$$

with some coefficients b_K. Using theorem 6.4 and the equation 6.29 we can compute the convolution of I with F:

$$I \star F = \sum_K b_K X_K \star F = \sum_K b_K c_K(F) X_K \tag{6.30}$$

where the constants c_K are given by the integrals in equations 6.28 and 6.29. We now apply the Laplacian and use theorem 6.3 to calculate the Laplacian of the filtered image:

$$
\begin{aligned}
E_F &= \sum_K a_K X_K \\
&= \nabla^2 (I \star F) \\
&= \nabla^2 \left(\sum_K b_K c_K(F) X_K \right) \\
&= \sum_K b_K c_K(F) \nabla^2 X_K \\
&= \sum_K b_K c_K(F) \lambda_K X_K
\end{aligned}
$$

From the equations 6.31 we find the relation between the coefficients of E and I:

$$a_K = b_K \cdot c_K(F) \cdot \lambda_K \tag{6.31}$$

The reader should have no difficulty in finding restrictions on F and I which make all the manipulations above possible (see also [50] for a set of conditions on F).

6.4   Image Restoration and Tomography

Image restoration is one application where one can use eigenspace decomposition of a given Hilbert space. From Fourier optics (see for example [16] and [38]) it is known that (under certain conditions) optical systems can be described as linear systems. If $o(x, y)$

describes the object function and $i(x, y)$ the observed image then we have the following relation between object and image:

$$i = o \star t. \tag{6.32}$$

Here $\star$ denotes the convolution and the function t describes the properties of the imaging system. Taking Fourier transforms this becomes:

$$I = O \cdot T \tag{6.33}$$

In many cases, for example for systems with circular aperture, t is only a function of the radius $r = \sqrt{x^2 + y^2}$. In these cases the equation 6.33 becomes:

$$I(\rho, \phi) = O(\rho, \phi) \cdot T(\rho)$$

and we see immediately that the operator K defined as

$$Ko = o \star t$$

intertwines with the group $SO(2)$.

The eigenfunctions of K are those object distributions that pass the system without distortion. For group theoretical reasons these eigenfunctions are related to the complex exponentials.

We use this information to recover the unknown object function o from the known image function i and the known transfer function t :

As before we expand the object function o and the image function i into a series of complex exponentials

$$o(r, \phi) = \sum_n g_n(r)e^{in\phi}$$

$$i(r, \phi) = \sum_n h_n(r)e^{in\phi}.$$

and apply the operator K to o. This gives for the measured image i :

$$\sum_n h_n(r)e^{in\phi} = i(r, \phi) = Ko = \sum_n \tilde{h}_n(r)e^{in\phi}.$$

The transformed radial weight function $\tilde{h}_n$ is given by:

$$\tilde{h}_n(r) = \tilde{K}(g_n)(r)$$

where:

$$K(g_n(r)e^{in\phi}) = \tilde{K}(g_n(r))e^{in\phi}.$$

since the complex exponentials are eigenfunctions of K. From this equation we can compute the unknown functions g_n if we can invert the operator $\tilde{K}$. Again we reduce the two-dimensional problem of inverting the operator K into the one-dimensional problem of inverting $\tilde{K}$.

The rotation invariance of the operator K was used to find the right type of basis functions $h_m(r)e^{im\phi}$ in the series expansion of the object function.

Another application of series-expansion type of reconstruction methods can be found in the area of CT-reconstruction. We will here only develop the main idea (see [9] for more details).

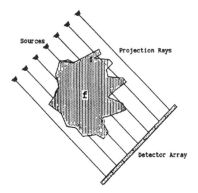

Figure 6.1: Principle of an X-ray Computer Tomograph

The simplest X-ray tomographs consist of a number of X-ray sources and a number of detectors in an arrangement of the type shown in figure 6.4.

The domain $\mathcal{D}$ on which the object functions have non-zero values is given by the unit disc. The (unknown) object function that has to be recovered is given by f. For each projection ray L we measure the remaining intensity of the X-ray when it has passed the object. We denote this intensity by $p(L)$. The line L can be described as $L = \{x \,|\, \langle x, \xi \rangle = c\}$ where c is a scalar variable describing the distance of the line to the origin and where ξ is a unit vector describing the direction of the projection ray. The equations $\langle x, \xi \rangle = c$ and $\langle x, -\xi \rangle = -c$ obviously describe the same line and we will therefore describe the projection rays with arbitrary unit vectors ξ and positive constants c. Since unit vectors in 2-D are uniquely specified by an angular variable ψ we see that L depends on the two parameters $c \geq 0$ and $\psi \in [0, 2\pi)$. Instead of writing $p(L)$ we can also describe the measurement by the function $p(c, \psi)$. The operation of a tomograph is described by an operator K that maps the original density distribution f into the measurement function p.

K intertwines with $SO(2)$ since a rotated object leads to a rotated projection:

$$[Kf(r, \phi + \alpha)](\rho, \psi) = [Kf(r, \phi)](\rho, \psi + \alpha).$$

Again we develop the original distributions f and p into series of circular harmonic functions and get for the original distribution f and the projections the expansions

$$\begin{aligned} f(r, \phi) &= \sum g_n(r) e^{in\phi} \\ p(r, \phi) &= \sum h_n(r) e^{in\phi}. \end{aligned}$$

Applying the projection operator K gives

$$\begin{aligned} \sum h_n(r) e^{in\phi} &= p(r, \phi) = Kf(r, \phi) \\ &= K(\sum g_n(r) e^{in\phi}) = \sum \tilde{K}(g_n(r)) e^{in\phi}. \end{aligned}$$

or $h_n(r) = \tilde{K}(g_n(r))$ for all n. Reconstructing the original distribution o from the measurements p amounts thus to an inversion of the operator $\tilde{K}$.

The new operator $\tilde{K}$ is given by:

$$\tilde{K}(g_n)(r) = 2 \int_p^\infty g_n(r) T_n \left(\frac{p}{r}\right) \left(1 - \frac{p^2}{r^2}\right)^{-1/2} dr$$

The two-dimensional problem of inverting the operator K was thus transformed to the one-dimensional problem of inverting $\tilde{K}$.

For more information on series expansion methods in CT-tomography the reader may consult the literature, for example [9].

The previous theory can be generalized to the n-dimensional case where one reconstructs the original distribution from a set of line integrals. This is the so-called X-ray transform. A detailed investigation of the inversion of the X-ray transform in n-dimensions using series expansions can be found in [21].

6.5 Filter Design

Feature extraction is one of the most important tasks in pattern recognition. In this section we will investigate a very general model of a feature extraction unit. Basically we assume that such a unit transforms an input signal s into a response signal $f(s)$. We call $f(s)$ the feature extracted from the input signal s. We can think of the input signal s as a part of an image detected by a camera or the retina. The feature extraction unit is then a cell or a pattern recognition algorithm that analyzes the incoming gray-value pattern. The computed result $f(s)$ is a description of the incoming pattern s. In the case where we think of the feature extraction unit as a cell, we could say that the incoming signal s describes the output of the cells in the receptive field of the feature extraction cell. The output $f(s)$ is the reaction of the feature extraction cell to the incoming signal. The purpose of this discussion, however, is not to describe natural cells but to develop a mathematical model that is based on mathematically motivated criteria.

The feature extraction unit is obviously completely specified if it is known how the response f depends on the input s. Designing such a unit is equivalent to constructing the mapping f. Many approaches to designing such units use an ad hoc strategy depending on the problem at hand. Others study natural systems, like the human vision system, to find out how well working systems have solved the problem. Many of these approaches have in common that they treat the set of input signals as just an unstructured set. In reality, however, we find that the space of input signals is often highly structured in the sense that the space can be partitioned into different subspaces where each subspace consists of essentially the same signals. In this paper we assume that the signal space can be partitioned into subspaces in such a way that all the signals in a subspace are connected to one another with the help of a group-theoretically defined transformation. We will also say that the subspaces are invariant under the group operation, or that the signal space is symmetrical.

As an example consider a feature extraction unit analyzing the gray value distribution in the neighborhood of a point in an image. Assume further that it is the goal of this unit to find out if the incoming gray value pattern was edge-like or not. We will refer to this problem as the (2-D) edge-detection problem. In the following we will denote by s_v, s_h and s_n the gray value distribution of a vertical edge, a horizontal edge and a noise pattern respectively. In the case where we assumed that the signal space is unstructured we would treat s_v, s_h and s_n as just three input patterns. We would completely ignore the additional information that s_v and s_h have more in common then s_v and s_n. In contrast to this unstructured approach we will in the following describe a method that is based on the observation that an effective feature extraction unit should exploit the additional information. In this theory we divide the space of input signals into subspaces of "essentially the same signals." We call such a subspace an invariant subspace. In the edge-detection example one such subspace would consist of all rotated versions of a fixed gray value distribution. Now s_v and s_h lie in the same subspace whereas s_n lies in another subspace. We can thus say that the similarity between s_v and s_h is greater than the similarity between s_v and s_n.

The purpose of this section is to show how to construct a feature extraction unit that can exploit the regularity of the space of input signals. We show that such a unit computes two "essentially equal" feature values $f(s_1)$ and $f(s_2)$ for two "essentially equal" signals s_1 and s_2. We demonstrate also that the unit can recognize all signals in a given invariant subspace once it has seen one member of it. In the edge detection example mentioned above we could say that such a unit can recognize a whole class of patterns, the class of edge-like gray value distributions. Furthermore the system can recognize all edges once it has encountered one special example of an edge.

In the next section we will develop the concept of a symmetrical signal space. Then we will use some results from representation theory to get an overview of all linear feature extraction units on symmetric signal spaces. We will also derive some important properties of these feature extraction units. We will illustrate the general theory with an example where the signals are gray value distributions and where the regularities or symmetries are defined with the help of the group of 2-D rotations.

6.5.1   Regularities in the Signal Space

In the edge detection example we saw that that we would like to introduce some type of structure in the space of all possible input signals. Now we will introduce a method that allows us to divide the space of all possible input patterns into subspaces of essentially the same patterns.

When we develop our model of a regular or a symmetrical signal space it might be useful to have a concrete example in mind against which we can compare the general definitions and constructs. We will mainly use the familiar 2-D edge-detection problem as such an example. The receptive field of the feature extraction unit consists in this case of a number of detectors located on a disk. If x denotes a point on this disk then we denote by $s(x)$ the intensity measured by the detector located at x. The input to the unit, the signal s, is in this example the light intensity measured by the sensors on the disk. Since the detectors are located on a disk we can assume that s is a function defined on some disk. For notational simplicity we can assume that this disk has the radius one and we will also assume that the function s is square-integrable. We denote the unit disk by $\mathcal{D}$ and we assume that our signal space is $L^2(\mathcal{D})$, the space of all square-integrable functions on the unit disk. Two typical step-edges from this space are shown in figure 6.2.

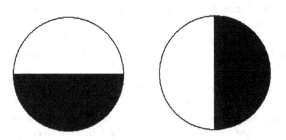

Figure 6.2: Two orthogonal, equivalent gray value distributions

From functional analysis we know that this signal space forms a Hilbert space, i.e. there is a scalar product $\langle\,,\,\rangle$ with which we can measure the length of a signal and the cosine of the angle between two signals as $\|s\|^2 = \langle s, s\rangle$ and $\|s_1\|\,\|s_2\|\cos(s_1, s_2) = \langle s_1, s_2\rangle$. Furthermore we can compute the distance between two signals as $\|s_1 - s_2\|$. This is about all we can say about the signals as elements of the Hilbert space. In our case these measurements can be quite misleading as a look at figure 6.2 shows. As elements of the Hilbert space these two signals are orthogonal, i.e. as uncorrelated or different as possible. From a higher level of understanding, however, they are nearly identical; they are only rotated relative to one another. We thus have to add a higher level of similarity to our basic, low-level Hilbert space model.

In the edge-detection problem we introduce such a similarity by defining that two gray value distributions are essentially equal if they are rotated versions of each other. In this way we get a higher level symmetry or a regularity defined by the set of all rotations. In this section we show how to build a model that describes such a regular signal space. We will then investigate what can be said about feature extraction units that can exploit these regularities.

We translate our basic concepts from group theory to pattern recognition language as follows:

Definition 6.1   1. A *transformation* of a space X is a continuous, linear, one-to-one mapping of X onto itself. The set of all transformations of X is a group under the usual composition of mappings and it is denoted by $GL(X)$.

2. A *regular* or a *symmetrical signal space* is a triple (H, G, T) of a Hilbert space H, a compact group G and a representation $T : G \to GL(H)$. The Hilbert space H is called the *signal space*, the group G is called the *symmetry group* of the symmetrical signal space and the mapping T is called its *representation* .

3. If $g \in G$ is a fixed group element and $s \in H$ is a fixed signal then we say that $T(g)s$ is the *transformed version* of s. Sometimes we will also write s^g instead of $T(g)s$.

In our edge detection example these definitions translate as follows: The Hilbert space H is the signal space $L_2(\mathcal{D})$ of square integrable intensity distributions on the unit disk. If

$s \in H$ is a given intensity distribution and g is a 2-D rotation then we can map this signal $s(x,y)$ onto the rotated intensity distribution $s(g^{-1}(x,y))$. Now we select a fixed rotation g. The mapping $T(g)$ that maps a function $s \in H$ to its rotated version is obviously a linear mapping from H into H. The symmetry group G is in this example the group of 2-D rotations, denoted by $SO(2)$. The mapping T maps the rotation g to the rotation operator $T(g)$ where $(T(g)s)(x,y) = s(g^{-1}(x,y))$. Note that $T(g_1 g_2)s = s((g_1 g_2)^{-1}) = s(g_2^{-1} g_1^{-1}) = T(g_1)(s(g_2^{-1})) = (T(g_1)T(g_2))s$ which explains the use of the inverse of g in the definition of T.

Our definition of a symmetrical signal space is slightly more restrictive then necessary. For most of our results it would have been sufficient to require that the group G is locally compact instead of compact. The resulting theory is however technically more complicated and therefore we will restrict us in the following to compact groups. Among the possible symmetrical signal spaces there are of course also uninteresting ones. As an example take $L^2(\mathcal{D})$ as the Hilbert space and $G = SO(2)$ as above. As representation select the mapping $g \mapsto id$ where id is the identity in $GL(H)$. For every signal $s \in H$ and all $g \in G$ we have thus $T(g)s = s$, the only transformed version of the signal is the signal itself. This example demonstrates that the mapping T is responsible for the power of the constraints the symmetry group has on the signal space: In the case where all group elements are mapped to one element, the identity, we have no constraint at all, the pattern classes consist of one element only. If the mapping T is such that for a fixed s the set $\{T(g)s | g \in G\}$ is a large set of signals then the constraints imposed by the group are very powerful.

We now define exactly what we mean when we speak of "essentially the same" signals. If (H, G, T) is a symmetrical signal space then we define two signals $s_1, s_2 \in H$ as *GT-equivalent* if there is a $g \in G$ such that $T(g)s_1 = s_2$, in this case we write $s_2 = s_1^g$ or $s_1 \overset{GT}{=} s_2$. If p is an element in H then we denote the set of all signals $s \in H$ that are equivalent to p by $p^G = \{s \in H | s = p^g; g \in G\}$. We call p^G a *pattern class* and p the *prototype pattern* of p^G. The set of all pattern classes in H will be denoted by H/G. By abuse of language we will also denote a complete set of prototype patterns by H/G the reader should have no difficulty in understanding what we are referring to.

From the properties of T it is clear that GT-equivalence is an equivalence relation and H can thus be partitioned into pattern classes. This means that H is the disjoint union of the different pattern classes, i.e. $H = \bigcup_{H/G} p^G$ and if $s_1^G \cap s_2^G \neq \emptyset$ then $s_1^G = s_2^G$. These definitions describe the high level similarity that we were referring to earlier: two signals represent essentially the same pattern if they lie in the same pattern class, independent of their Hilbert space relation.

The simplest illustration of the group theoretical results developed so far is provided by the symmetrical signal space (H, G, T) where H is the space of all square integrable functions on the unit circle, $G = SO(2)$ and where T is the rotation operator $T(g)s(\phi) = T(\psi)s(\phi) = s(\phi - \psi)$ (ψ is the rotation angle of the rotation g). An element $s \in H$ can be decomposed into a Fourier series $s(\phi) = \sum c_n e^{2\pi i n \phi}$, i.e. H is the direct sum of the one-dimensional subspaces H^n consisting of the complex multiples of $e^{2\pi i n \phi}$. The representation T can be reduced to the sum of one-dimensional, irreducible representations T^k defined by $T^k(g)s(\phi) = T^k(\psi)(\sum c_n e^{2\pi i n \phi}) = \sum_{n \neq k} c_n e^{2\pi i n \phi} + c_k(e^{2\pi i k(\phi - \psi)})$. The (1-D) matrix representation is given by the matrix entry $t_{11}^k(g)$ defined by $g \mapsto e^{-2\pi i k \psi}$.

6.5.2   Application to Feature Extraction

We apply now the results presented so far to investigate feature extraction units. For this purpose we assume that the signal space of our unit is a symmetrical signal space (H, G, T) as defined in definition 6.1. In the discussion of this definition we noticed that GT-equivalence is an equivalence relation and we partitioned H into the equivalence classes H/G. As a result we can now decompose an arbitrary signal $s \in H$ into a component depending on G and one depending on the prototype: we write $s = s(g, p)$. We now define what a feature is:

Definition 6.2   1. A *feature* is a complex valued function f on H. If we write $f(s) = f(s(g, p)) = f(g, p)$ then we will also assume that f, as a function of g, is square-integrable on G.

2. A *linear feature* is a feature which is also a linear map, i.e. it satisfies the condition: $f(c_1 \cdot s_1 + c_2 \cdot s_2) = c_1 \cdot f(s_1) + c_2 \cdot f(s_2)$ for all signals $s_1, s_2 \in H$ and all complex constants c_1, c_2.

With the help of the Peter-Weyl theorem we can now give a simple description of features: if we keep the prototype signal $p \in H/G$ fixed, then f is a function of g and we can use the Peter-Weyl theorem to get a series expansion:

$$f(g, p) = \sum_k \sum_{nm} \alpha_{nm}^k(p) t_{nm}^k(g) \tag{6.34}$$

This describes how f depends on p and g. We may also say that we have separated the dependencies of f on p and g.

Using this expression for the feature f and the orthogonality of the functions $t_{nm}^k(g)$ we can apply Haar integration to compute the mean value of f over one pattern class or the correlation between the signals in two pattern classes. For the mean we get: $\int_G f(g, p)\, dg = \alpha_0(p)$ where α_0 is the coefficient belonging to the trivial, 1-dimensional representation defined by $t(g) = 1$ for all $g \in G$. For the correlation we get:

$$
\begin{aligned}
\int_G f(g, p_1) f(g, p_2)\, dg &= \int_G \left(\sum_k \sum_{n,m} \alpha_{nm}^k(p_1) t_{nm}^k(g) \right) \left(\sum_l \sum_{i,j} \alpha_{ij}^l(p_2) t_{ij}^l(g) \right) dg \\
&= \sum_k \sum_{n,m} \sum_l \sum_{i,j} \alpha_{nm}^k(p_1) \alpha_{ij}^l(p_2) \int_G t_{ij}^l(g) t_{nm}^k(g)\, dg \\
&= \sum_k \sum_{nm} \alpha_{nm}^k(p_1) \alpha_{nm}^k(p_2) d_k^{-1}
\end{aligned}
$$

For our purposes these features are too general and in the rest of this section we will therefore only consider linear features and simple functions of linear features.

Assume now that (H, G, T) is a symmetrical signal space and that the representation T is the direct sum of a complete set of inequivalent, unitary, irreducible representations T^k of G. We denote the Hilbert space belonging to the irreducible, unitary representation T^k by H^k. The dimension of this space is denoted by d_k and the basis elements of H^k are the signals $e_1^k, \ldots, e_{d_k}^k$. The matrix entries of T^k with respect to this basis are the functions $t_{nm}^k(g)$. For a fixed index k we denote the matrix of size $d_k \times d_k$ and elements $t_{nm}^k(g)$ by $T^k(g)$. For simplicity we will also assume that H is the direct sum of all the subspaces H^k. A basis of H is thus given by the elements e_n^k.

Using the fact that the elements e_n^k form a basis of H we find for a prototype signal p the following expansion:

$$p = \sum_k \sum_{n=1}^{d_k} \beta_n^k e_n^k. \tag{6.35}$$

Using the scalar product in the Hilbert space H we can compute the coefficients β_n^k as usual by $\beta_n^k = <p, e_n^k>$. The subspaces H^k are invariant under the group operations and the transformation of the basis elements is described by the matrices $T^k(g) = (t_{nm}^k(g))$. For an arbitrary signal $s = s(g, p) = p^g$ we find therefore the expansion:

$$s = p^g = \sum_k \sum_{nm} \beta_m^k t_{nm}^k(g) e_n^k. \tag{6.36}$$

From this expansion and the linearity of the feature function f we get the following equation for the value of the linear feature f:

$$f(s) = f(p^g) = \sum_k \sum_{nm} \beta_m^k t_{nm}^k(g) f(e_n^k). \tag{6.37}$$

This equation shows how f depends on the prototype signal p, the group element g and the value of f on the basis elements e_n^k. We see also that a linear feature function f is completely defined by its values at the basis elements e_n^k. Among the linear features there is a set of especially simple ones which will be defined in the next definition:

Definition 6.3   1. A linear feature which has the value 1 at one fixed basis element e_n^k and which is 0 for all other basis elements is called a *basic linear feature*. It will be denoted by f_n^k.

2. A *basic feature vector* f^k is a map $f^k : H \to \mathbf{C}^{d_k}$ of the form $s \mapsto f^k(s) = (f_1^k(s), ..., f_{d_k}^k(s))$ where the f_n^k are the basic features belonging to the subspace H^k of the representations T^k.

Equation 6.37 shows that the all linear features are linear combinations of basic linear features and in the rest of this paper we will therefore restrict our attention to basic linear features.

For the prototype signal p we get from the equations 6.35 and 6.37 the following feature value: $f(p) = \sum_k \sum_n \beta_n^k f(e_n^k)$. If f is a basic linear feature, say f_ν^μ then we have:

$$f_\nu^\mu(p) = \beta_\nu^\mu. \tag{6.38}$$

If s is a signal which is GT-equivalent to $p : s = p^g$ then we find for the feature value (see equation 6.37):

$$f_\nu^\mu(s) = f_\nu^\mu(p^g) = \sum_k \sum_{nm} \beta_m^k t_{nm}^k(g) f_\nu^\mu(e_n^k) = \sum_m \beta_m^\mu t_{\nu m}^\mu(g) = \sum_m f_m^\mu(p) t_{\nu m}^\mu(g) \tag{6.39}$$

Using the basic feature vector notation we find then the following theorem:

Theorem 6.6   1. $f^\mu(s) = f^\mu(p^g) = T^\mu(g) f^\mu(p)$

2. Since T^μ was a unitary representation we have for all $g \in G$:

$$\|f^\mu(p)\| = \|f^\mu(p^g)\| \tag{6.40}$$

The magnitude of the feature vector is thus invariant under the transformation $p \mapsto p^g$.

The previous results show that we can treat the basic feature vectors f^{k_1} and f^{k_2} separately if $k_1 \neq k_2$, but that we have to consider all basic linear features belonging to the same representation simultaneously. The equation 6.40 is of course only valid for basic feature vectors; if the different basis elements have different feature values then in general we will not have the same magnitude for all feature vectors of GT-equivalent signals.

The results in theorem 6.6 suggest the following procedure for our pattern recognition problem to detect a certain class of signals.

Assume we know that our signal space is symmetric and assume further that we know the symmetry group of our problem and the irreducible unitary representations of this group. In the learning phase we present the system with one prototype signal p. The feature extraction unit computes from this signal the feature vector

$$f^k(p) = (f_1^k(p), ..., f_{d_k}^k(p)) = (< p, e_1^k >, ..., < p, e_{d_k}^k >)$$

where the e_n^k form a basis of the subspace H^k belonging to the irreducible unitary representation T^k. The next time the unit receives an unknown signal s at its input it computes

$$f^k(s) = (f_1^k(s), ..., f_{d_k}^k(s)) = (< s, e_1^k >, ..., < s, e_{d_k}^k >).$$

Then it compares $\|f^k(p)\|$ and $\|f^k(s)\|$ and if these two values are different it concludes that the two signals do not belong to the same class. If they are more or less equal then we can conclude that they might be equivalent, i.e. that there might be a group element $g \in G$ such that $s = p^g$. But there is no guarantee that they are similar. If it is necessary to find the transformation g also then one might try to recover g from the matrix equation $f^k(s) = T^k(g)f^k(p)$. However, this is not always possible, as we will see, a little later, in an example.

We will now demonstrate how we can use the results obtained so far to get an overview of the symmetrical signal spaces and we will also show how to construct filter functions.

Consider a fixed signal space H and a fixed (compact) symmetry group G. If (H, G, T) is a symmetrical signal space then we know from the previous theorems that T is equivalent to the direct sum of finite-dimensional, irreducible representations T^k. This gives all the possible spaces (H, G, T). We saw earlier that the properties of T describe the strength of the symmetry constraints: if $T(g)$ is the identity mapping for all $g \in G$ then we impose no group theoretical constraint since every signal $s \in H$ is GT-equivalent only to itself. On the other hand if T is the direct sum of a complete set of irreducible representations of G then T imposes the strongest constraint possible.

Let us now return to our 2-D edge-detection problem. The signal space is in this case the space of all square-integrable functions defined on the unit disk $\mathcal{D}$: $H = L_2(\mathcal{D})$. The symmetry group of our problem is the group of 2-D rotations, which means that we want to detect the pattern independent of its orientation in the image. This group is commutative, and the irreducible unitary representations are therefore all one-dimensional

(Schur's lemma 3.4). If ψ is the rotation angle of the rotation g then we saw that the matrix entry $t_{11}^k(g)$ belonging to the irreducible representation T^k is the function $e^{-2\pi ik\psi}$. The Peter-Weyl theorem says in this case that functions $t_{11}^k(g) = e^{-2\pi ik\psi}$ form a complete function set on the unit circle. We see thus that the Peter-Weyl theorem is a generalization of the Fourier expansion of periodic functions on a finite interval. The one-dimensional subspaces H^k of H belonging to T^k are spanned by functions of the form $W_k(r)e^{2\pi ik\phi}$ with a radial weight function W_k. These so-called rotation-invariant operators were introduced in [6] and further investigated in a number of papers (for example [51], [52], [8],[1], [42], [27], [32], [2], [28], [29], [7], [30], [31]).

Now suppose that T^k is an irreducible representation of $SO(2)$. Then T^k transforms a signal s by multiplying the k-th term in the Fourier expansion with the factor $e^{-2\pi ik\psi}$. If the signal $s = s(r,\phi)$ has the Fourier decomposition $s(r,\phi) = \sum_l w_l(r)e^{2\pi il\phi}$ then the transformed pattern $T^k(g)s$ becomes:

$$T^k(g)s = T^k(\psi)s(r,\phi) = \sum_{l \neq k} w_l(r)e^{2\pi il\phi} + w_k(r)e^{2\pi ik(\phi-\psi)}. \qquad (6.41)$$

From this we see that two signals are GT^k-equivalent if their Fourier coefficients are equal for all $l \neq k$ and if the Fourier coefficients for the index k have the same magnitude. An arbitrary representation is the direct sum of irreducible, one-dimensional representations and two signals are equivalent if their Fourier coefficients are equal for all indices that are not involved in the representation. For all indices involved in the decomposition of the representation the Fourier coefficients have equal magnitude. The number of Fourier components involved in the representation T describes how powerful the symmetry constraint imposed by the representation T is.

The scalar product in H is given by the integral

$$< s_1(r,\phi), s_2(r,\phi) >= \pi^{-1} \int_0^1 \int_0^{2\pi} s_1(r,\phi)s_2(r,\phi)r \; d\phi dr.$$

If $W_k(r)e^{-2\pi ik\psi}$ is a fixed basis element e_1^k and if the signal s has the Fourier decomposition $s(r,\phi) = \sum_l w_l(r)e^{2\pi il\phi}$ then the computed feature value becomes: $< s, e_1^k >= \int_0^1 W_k(r)w_k(r)r \; dr$. Feature extraction amounts thus to a convolution of the signal function s with the filter function $W_k(r)e^{2\pi ik\phi}$ or to a Fourier decomposition of the signal in polar coordinates.

From the equation 6.41 we see also that two different group elements $g_1 \neq g_2$ can produce the same transformation $T^k(g_1) = T^k(g_2)$, we only have to take as g_1 the rotation with angle ψ and as g_2 the rotation with angle $\psi + 2\pi/k$. In general, therefore, we cannot recover the group element g from the feature values.

6.5.3   Optimal Basis Functions

We will now describe an approach to filter design that is, at first sight, totally different from the previously developed strategy. At the end of this subsection we will have shown, however, that also this filter design method leads to the same type of filter functions. This filter design strategy was proposed by Hummel and Zucker (see [20] and [56]) as a tool to construct two- and three-dimensional edge-detectors.

In the following we assume again that the input signals arriving at the feature extraction unit are elements of some Hilbert space H and that the feature extraction unit is linear. From the Riesz-representation theorem (see 2.8) we know that the unit can be described by a fixed element f of the Hilbert space. This element will be called a filter function. The feature extracted from the pattern p is given by $\langle p, f \rangle$. Also in this section we assume that we have a group G that operates on the Hilbert space. If we have a pattern p and a group element $g \in G$ then we denote the transformed pattern again by p^g.

The Hilbert space H is in practically all cases infinite-dimensional, i.e. there are infinitely many independent filter functions f^k. In a real application, however, we can only compute finitely many feature values. Therefore we have to select a finite number of these basic features. This problem cannot be solved by using symmetry considerations alone, instead we have to introduce additional constraints to select a finite number of filter functions. In the optimal basis function approach to filter design one defines an optimal filter function, or an optimal basis function, as the function f that correlates best with the whole class of functions $p^G = \{p^g : g \in G\}$. This leads us to the problem of finding the function f that is a solution of

$$\int_G |\langle p^g, f \rangle|^2 \, d\mu \overset{!}{=} \max \tag{6.42}$$

We investigate this problem under the following symmetry constraints: we assume that $\mathcal{D}$ is a region in some space, that G is a group and that the Hilbert space H is the L^2 space of square-integrable functions on $\mathcal{D}$. G acts on $L^2(\mathcal{KD})$ by $f(X) \mapsto f(g^{-1}X)$ as usual. We also assume that we have a scalar product of the form:

$$\langle p, f \rangle = \int_{\mathcal{D}} p(X) \overline{f(X)} \, d\mu(X)$$

and that the representation defined by $T : g \to T(g)$ with $T(g)f = f^g$ and $f^g(X) = f(g^{-1}(X))$ is unitary, i.e.

$$\int_{\mathcal{D}} p(gX) \overline{f(gX)} \, d\mu(X) = \int_{\mathcal{D}} p(X) \overline{f(X)} \, d\mu(X)$$

for all $g \in G$. Defining the function $k(X, Y)$ as

$$k(X, Y) = \int_G p^g(X) p^g(Y) \, dg$$

we get for the integral in equation 6.42

$$\begin{aligned}
\int_G |\langle p^g, f \rangle|^2 \, dg &= \int_G \left(\int_{\mathcal{D}} p^g(X) \overline{f(X)} \, d\mu(X) \int_{\mathcal{D}} f(Y) \overline{p^g(Y)} \, d\mu(Y) \right) dg \\
&= \int_{\mathcal{D}} \int_{\mathcal{D}} k(X, Y) f(Y) \overline{f(X)} \, d\mu(Y) d\mu(X).
\end{aligned} \tag{6.43}$$

Defining the operator $A : f \to \int_{\mathcal{D}} k(X, Y) f(Y) \, d\mu(Y)$ we find that equation 6.42 is equivalent to

$$\langle Af, f \rangle \overset{!}{=} \max \tag{6.44}$$

From functional analysis (see for example [53]) it is known that the solution of 6.44 exists and is equal to an eigenfunction of A if A is compact and symmetric. Again we are led to an eigenproblem. If we know that A is intertwining then we know that the eigenfunctions of A define a representation. In the standard example of 2-D rotation-invariant pattern recognition we find as optimal basis functions the filter functions of the type $h_n(r)e^{in\phi}$. For three-dimensional images one gets the rotation-invariant operators $h_n(r)Y_n^m(\phi,\theta)$. For a detailed treatment see [28].

6.5.4 An Ergodic Property

In physics one is often interested in ergodic processes. These processes have the property that the spatial average over some quantity is equal to the time average of the same quantity (see [38], [4] and [46] and the references cited there). In this section we will derive a general relation between weighted group averages and weighted spatial averages. This relation can easily be derived with the Peter-Weyl theorem.

We assume again that the pattern functions p are defined on the homogeneous space X (all the results can also be derived for the case where the space is of the form $R \times X$ with a homogeneous space X). We write thus $p = p(g), g \in G$. We assume also that we want to detect the signals of the form $p_0^g(h) = p_0(hg)$ (where p_0 is the fixed prototype pattern and g is a group element). We denote the matrix entries of a complete set of irreducible representations of G again by $t_n^{kl}(g)$. From the Peter-Weyl theorem it follows that the prototype pattern p_0 can be written as a series:

$$p_0(g) = \sum_{nkl} a_n^{kl} t_n^{kl}(g) \tag{6.45}$$

We compute now the spatial weighted average

$$\sigma_\nu^{\mu\kappa} = \int_G p_0(g) t_\nu^{\mu\kappa}(g)\, dg = \sum_{nkl} a_n^{kl} \int_G t_n^{kl}(g) t_\nu^{\mu\kappa}(g)\, dg = \frac{1}{N} a_\nu^{\mu\kappa} \tag{6.46}$$

where $\int_G \ldots dg$ is the Haar integral as usual and N is the dimension of the representation T_ν. From the definition of the transformed pattern $p_0^g(h) = p_0(hg)$ we find at once that this spatial average is equal to the following group average:

$$\sigma_\nu^{\mu\kappa} = \int_G p_0^g(e) t_\nu^{\mu\kappa}(g)\, dg = \int_G p_0(g) t_\nu^{\mu\kappa}(g)\, dg \tag{6.47}$$

where e is the identity element of the group G.

The spatial average and the weighted group average (at the identity) are thus equal. Using the orthogonality relations for the matrix entries (equations 5.3) we can easily derive a more general relation between spatial averages and group averages:

$$\int_G p_0^g(h) t_\nu^{\mu\kappa}(g)\, dg = \int_G p_0(hg) t_\nu^{\mu\kappa}(g)\, dg = \int_G p_0(g) t_\nu^{\mu\kappa}(h^{-1}g)\, dg \tag{6.48}$$

The functions $t_\nu^{\mu\kappa}(g)$ are the matrix entries of the representation $T_\nu(g)$ and these matrices transform as

$$T_\nu(gh) = T_\nu(g)T_\nu(h). \tag{6.49}$$

Combining the equations 6.48 and 6.49 we see at once that the weighted group average at the point h is a linear combination of the weighted spatial averages $\sigma_\nu^{\mu\kappa}$. The coefficients in this combination are given by the matrix entries $t^{\mu\kappa}_\nu(h^{-1}g)$. This result generalizes the result from section 1.3.

The previous discussion shows that it is possible to compute feature values that characterize the whole class of patterns (the weighted group averages) from one member of the pattern class (weighted spatial averages). It shows also that these features characterize the pattern class completely.

Note also that the previous result was derived in a constructive way, i.e. we know exactly how to compute these features. We summarize the previous results in the following algorithm to detect patterns of a given pattern class:

1. Given the prototype pattern p_0 compute the weighted spatial averages $\sigma^{\mu\kappa}_\nu$ defined in equation 6.46. Compute these features for a finite number of indices $\nu_1, ..., \nu_N$ and all indices μ and κ that belong to the chosen indices ν_k. These average values define, up to matrix multiplications, also the weighted group averages.

2. For an unknown pattern p compute the same average values $\tilde{\sigma}^{\mu\kappa}_\nu$.

3. If there is a group element h such that the feature values $\tilde{\sigma}^{\mu\kappa}_\nu$ and $\sigma^{\mu\kappa}_\nu$ are connected via the transformation matrices $T_{\nu_k}(h)$ for all selected indices ν_k then decide that the unknown pattern p belongs to the same class as p_0.

6.5.5 Some Examples

In the previous subsections we derived a general procedure for group invariant pattern recognition and in this subsection we will illustrate this strategy with two examples.

In the first example we want to recognize two faces, independent of their orientation. The two prototype patterns are shown in figure 6.3.

Using the algorithm described above we compute in the first step the weighted spatial averages and get the two radial functions

$$h_1(r) = \int_0^{2\pi} p_1(r,\phi)\, d\phi \quad \text{and} \quad h_2(r) = \int_0^{2\pi} p_2(r,\phi)\, d\phi \quad (6.50)$$

where $p_1(r,\phi)$ and $p_2(r,\phi)$ describe the gray value distribution for the first and the second face respectively.

Next we apply an orthogonalization procedure to compute the radial weight function $\tilde{h}$ defined as:

$$\tilde{h}(r) = h_1(r) - \int_0^1 h_1(r)h_2(r)r\, dr h_2(r). \quad (6.51)$$

In scalar product notation this becomes:

$$\tilde{h} = h_1 - \langle h_1, h_2 \rangle h_2$$

and it can be easily seen that $\tilde{h}$ is the component of h_1 that is orthogonal to h_2. We get $\langle \tilde{h}, h_1 \rangle = 1 - \langle h_1, h_2 \rangle^2$ and $\langle \tilde{h}, h_2 \rangle = 0$.

If we want to classify an unknown face $p(r,\phi)$ then we compute first the spatial averages $h(r) = \int_0^{2\pi} p(r,\phi)\, d\phi$ and then

$$\int_0^t h(r)\tilde{h}(r)r\, dr = c. \quad (6.52)$$

Figure 6.3: Two prototype patterns

If the constant c is above a certain threshold then we decide that p was similar to p_1 otherwise we decide that it was similar to the face p_2.

In the next figure 6.4 we see a number of faces in different orientations and with different amounts of noise added. In figure 6.5 we have coded the filter result (computed with the formula 6.52) into a gray value and put a square with the corresponding gray-value in the middle of the input pattern. We see that all patterns were correctly classified.

In the next example we used the filter strategy to detect blood vessels in three-dimensional images generated by an MR-scanner. In our experiments we used a set of two 3-D images from an MR-scanner. Both images consisted of $256 \times 256 \times 32$ pixels each. Each pixel contained a 12-bit density value. These two images are used to study the blood flow in the knee-region. For a detailed description the reader may consult [26]. Today these images are evaluated by combining the gray values from the two different images by simple arithmetic operations. By pointwise subtracting the two images it is possible, for example, to extract all points belonging to the blood vessels.

In our experiments we computed the pixelwise difference between the two images and thresholded it. This thresholded image was used as our target pattern and we wanted to detect the blood structures by using the gray value information from one image only.

We filtered the first image with the three edge- and the five line-detection filters $Y_n^m (n = 1, 2)$. The filter kernels consisted of $5 \times 5 \times 5$ points. The feature vector computed for each point consisted of nine components: the gray value of the original image, the three edge features and the five line features.

$$\text{Feature Vector} = \begin{pmatrix} \text{original} \\ x \\ y \\ z \\ l1 \\ l2 \\ l3 \\ l4 \\ l5 \end{pmatrix}$$

Neural nets, especially the back propagation net, are interesting strategies for evaluating these feature vectors. Their advantages are trainability and good performance even for very complex decision regions. For more information on neural nets and back propagation procedures the user may consult the literature (see for example [33], [12] and [13]). The only problem left to the user is to choose a small training region. The selected region must provide representative selection of target and non-target points. Training of the nets by samples from the feature vectors usually converges very fast, i.e. in 300 iterations.

After the pixels in the image have been classified by the neural net the result can be enhanced by applying further image processing methods. In our study we used noise reduction algorithms to eliminate isolated, misclassified points.

An overview of the classification method used in this example is shown in figure 6.6. Experiments showed that the l3 feature is unnecessary and rather confuses the net. This might be a consequence of the different sampling of the 3-D image where the distance between two pixels in z-direction (between two slices) was larger than the distance in x- and

Figure 6.4: A number of transformed patterns

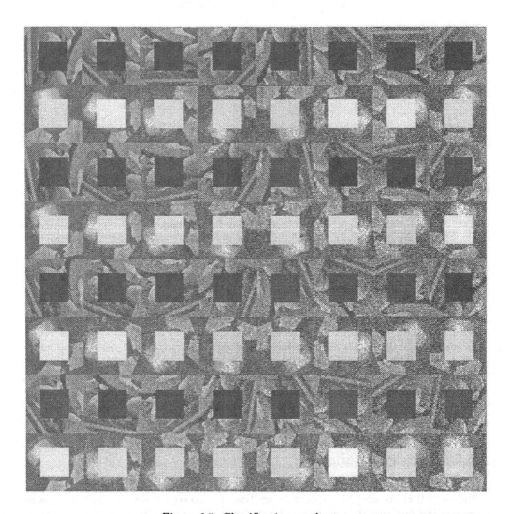

Figure 6.5: Classification results

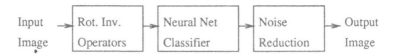

Figure 6.6: Structure Detection Procedure

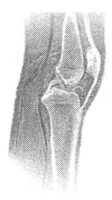

Figure 6.7: Slice 16 in Input Image

y-direction (between pixels in one slice). The filter kernels were designed to compensate this anisotropy but the information from different slices is of course not as reliable as the information from different pixels in one slice.

To train the neural net a user has to mark a few pixels belonging to blood vessels and some background pixels. The net which produced the output in figure 6.8 was trained with 13 pixels – or rather feature vectors. The training set must be a connected region to make sense as input to the net: one cannot pick 5 random blood vessel pixels and 5 random non-blood vessel pixels. Figure 6.7 shows one slice from the original input image. The blood vessels can be seen as dark vertical structures at the left side of the knee. Figure 6.8 shows the result after net classification and the final classification result after noise reduction is shown in figure 6.9. It is worth noting that merely thresholding the input image produces only poor results.

6.5.6   Image Coding

In the last sections we saw how we can use group theoretical methods to design filter functions that are useful in pattern recognition. In this section we reformulate the same results and find that they might also be applicable for image coding.

In many applications of digital image processing it is necessary to describe images in compressed form. A typical example is the transmission of images. In these applications one can often accept a certain degradation of image quality if one can thereby achieve a

Figure 6.8: Slice 16 in Output Image

Figure 6.9: Slice 16 in Output Image after Noise Reduction

shorter transmission time. In this section we briefly mention one image coding technique that is based on the so-called Karhunen-Loeve expansion.

This coding technique works as follows:

- In the first step the transmitter and the receiver agree upon using a certain set of functions for coding. These functions are denoted by $f_1, ..., f_n$.

- If an image is to be transmitted then the transmitter approximates the image function i with the sum $\tilde{i} = a_1 f_1 + ... + a_2 f_n$.

- Having calculated the coefficients a_i from the original image i the transmitter sends these coefficients to the receiver.

- The receiver restores the original image from the received signals a_i by computing $\tilde{i} = a_1 f_1 + ... + a_2 f_n$.

The efficiency of this process can be measured by comparing the number of bytes needed to transmit the original image i and the number of bytes needed to describe the coefficients a_i. The quality can be measured by computing the difference between i and $\tilde{i}$. If we fix the number n (i.e. number of coefficients = the number of the filter functions) then we can define an optimal coding process as one that minimizes the difference between i and $\tilde{i}$. If only one image is to be transmitted then we can of course compute the f_k from the image. In reality, however, we will mostly transmit a number of different images. In this case we have to find functions f_k that minimize the mean difference between the transmitted and the original images. A different way to express the same thing is to say that we want to find those functions f_k that correlate best with the set of images to be transmitted.

In a real application one does not code the whole image in one single step but one first partitions the original image into a number of subimages and then submits these subimages separately. This has the advantage that the subimages are often very homogeneous so that one needs only a few parameters to describe them.

This approach to image coding is identical to the optimal basis function approach to filter design. The only difference is that in the filter design problem we use the results for pattern recognition whereas in the image coding application we send the results to the receiver as a description of the original image. Both problems have in common that one has to find a number of functions that describe the information contained in the original image in an optimal way.

We mentioned earlier that the sender divides the original image into a number of subimages and that these subimages are often highly homogeneous. Very often these subimages do not contain any specific structures and we can therefore assume that there is a large group of transformations g with the property that the subimages i and i^g are equally probable in the input sequence. Examples of such a group are the rotation, translation and scaling groups.

If we know that our sequence of input images has this group theoretical homogeneity property then we conclude by the same arguments as in the section on optimal filter functions that the optimal set of functions f_k is given by the matrix entries of the irreducible group representations.

6.6 Motion Analysis

In this section we will introduce a group theoretically motivated model for motion analysis. This model generalizes the work of Kanatani on camera rotation reported in [22] and [23]. From a group theoretical point of view this model is interesting since it shows that that also non-compact groups, like $SL(2, \mathbf{C})$ might have interesting applications in image science. In the next subsection we introduce first the stereographic projection that will provide us with the connection between $M(3)$ and $SL(2, \mathbf{C})$.

6.6.1 Stereographic Projection

Recall that a 3-D rigid motion is defined as a rotation of 3-D space followed by a translation. Assume that (x, y, z) are the coordinates of a point P in some cartesian coordinate system and that P is mapped under the rigid motion m onto $P' = m(P)$. Then we find that the coordinates of $m(P)$ in the same coordinate system are given by the matrix equation

$$g \begin{pmatrix} x \\ y \\ z \end{pmatrix} + \begin{pmatrix} t_x \\ t_y \\ t_z \end{pmatrix} = g \cdot P + h \qquad (6.53)$$

where g is a 3×3 rotation matrix and h is a 3-D translation vector. By an easy calculation it can be shown that this form of the matrix equation is independent of the particular coordinate system selected. We will therefore first select a coordinate system that is especially well suited for our purposes. The Z-axis of this coordinate system is given by the optical axis of the system and the XY-plane contains the sensor array (the retina) of the system (see figure 6.10). We assume that the system has a circular aperture with center at the origin of the XYZ-coordinate system. The XY-coordinate axes are scaled so that this aperture describes a disc of radius $1/2$ in the XY-plane. The Z-axis is scaled so that the focal point F of the system has the coordinates $(0, 0, 1/2)$.

The XY-plane will be identified with the complex plane and in this plane we will identify the point with coordinates $(\xi, \eta, 0)$ with the complex point $\zeta = \xi + i\eta$. The intensity values measured by the sensor array can thus be described by a function $f(\zeta)$ that is defined on the unit disc $\mathcal{D} = \{\zeta | |\zeta| \leq 1\}$. f has positive real function values. If the point $\zeta = \xi + i\eta$ is a fixed point on the unit disc then we compute $f(\zeta)$ as follows: We draw a line from the focal point F through the sensor point $\zeta = \xi + i\eta = (\xi, \eta, 0)$. After the ray has left the sensor we detect where it hits the first visible object point. This point is denoted by $P(\zeta)$. We assume that the detected intensity value $f(\zeta)$ is given by the intensity value at the scene point $P(\zeta)$.

By a simple calculation we find the following relation between the coordinates (x, y, z) of the scene point $P(\zeta)$ and the coordinates (ξ, η) of its image point:

$$\xi = \tfrac{x}{1/2 - z} \quad \text{and} \quad \eta = \tfrac{y}{1/2 - z} \qquad (6.54)$$

It is easy to see that the stereographic projection maps the complex plane homeomorphically onto the unit sphere with the north pole deleted. Adding a new point, usually denoted by ∞, to the complex plane one gets a continuous one-to-one mapping between the whole sphere and the complex plane together with the new point. This new set consisting of the complex plane and the point ∞ is called the extended complex plane.

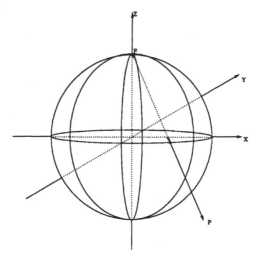

Figure 6.10: Stereographic Projection

6.6.2   The Effect of Rigid Motions

Before we investigate the effect of a rigid motion we will recall some facts from section 4.3:

Assume that a, b, c and d are four complex numbers with $ad - bc = 1$, then we denote the matrix $\begin{pmatrix} a & b \\ c & d \end{pmatrix}$ by $\sigma(a, b, c, d)$. We define also a function σ :

$$\sigma(a, b, c, d)\zeta = \tfrac{a\zeta + b}{c\zeta + d}$$

for a complex variable ζ. Note that $\sigma(a, b, c, d)$ denotes both a matrix and a function but it should be clear from the context what interpretation is meant. The group SL(2,C) acts on the extended complex plane via the transformations $\zeta \mapsto \tfrac{a\zeta + b}{c\zeta + d}$.

A simple calculation shows that the following relation holds for the combination of two such functions (matrices):

$$\sigma(a, b, c, d)\left(\sigma(\alpha, \beta, \gamma, \delta)\zeta\right) = \left(\sigma(a, b, c, d)\sigma(\alpha, \beta, \gamma, \delta)\right)\zeta \qquad (6.55)$$

The concatenation of two such functions is thus given by the function described by the multiplication of the two matrices of the original matrices.

In the next theorem we describe how rigid motions are mapped into SL(2,C):

Theorem 6.7 Denote by P the point with coordinates (x, y, z) and by $P^m = m(P)$ the image of P under the 3-D rigid motion m. Assume m is given by the transformation

$$P^m = g \cdot P + h = g(\varphi_1, \theta, \varphi_2)\begin{pmatrix} x \\ y \\ z \end{pmatrix} + \begin{pmatrix} t_x \\ t_y \\ t_z \end{pmatrix} \qquad (6.56)$$

where $g(\varphi_1, \theta, \varphi_2)$ is a 3-D rotation with Euler angles φ_1, θ and φ_2. The image of P under the stereographic projection is denoted by ζ and the image of P^m by ζ^m. Then we have the following relation between ζ and ζ^m :

$$
\begin{aligned}
\zeta^m &= \sigma(\delta,\gamma/\delta,0,1/\delta)\left(\sigma(\alpha,\beta,-\overline{\beta},\overline{\alpha})\zeta\right) \\
&= \left(\sigma(\delta,\gamma/\delta,0,1/\delta)\sigma(\alpha,\beta,-\overline{\beta},\overline{\alpha})\right)\zeta
\end{aligned}
\tag{6.57}
$$

To prove this theorem one uses equation 6.55. There we saw that the concatenation of two linear mappings in three-dimensional space leads to a matrix multiplication of the resulting matrices describing the fractional linear mappings. It suffices therefore to consider rotations and translations separately. Next one splits an arbitrary rotation g into a product of three rotations $g = g_{\varphi_1} g_\theta g_{\varphi_2}$ where φ_i and θ are the Euler angles of the rotation g (theorem 2.27). The first and the last of these rotations describes a rotation around the Z-axis and g_θ is a rotation with angle θ around the X-axis. Using equation 6.55 it is thus sufficient to investigate three cases:

- Translations

- Rotations around the Z-axis

- Rotations around the X-axis.

Translations are simple to deal with: one only has to note that translations along the X- and Y-axis lead to translations in the complex plane, whereas translations along the Z-axis gives a scaling. Both, translation and scaling, are described by matrices of the form $\sigma(\delta,\gamma/\delta,0,1/\delta)$. Rotations around the Z-axis are also easy since they result in rotations in the image plane. Such a rotation is described by the function $\zeta \mapsto e^{i\varphi}\zeta$.

This leaves the case of a rotation g_θ around the X-axis. We define w as

$$
w = \frac{y + iz}{1/2 - x}.
$$

By the same calculation as above we find that g_θ maps w into

$$
w' = g_\theta(w) = e^{i\theta}w.
\tag{6.58}
$$

Using the definition of w we find the following relation between w and

$$
\zeta = \frac{x + iy}{1/2 - z} :
$$

$$
\frac{w + i}{w - i} = \frac{\frac{y+iz}{1/2-x} + i}{\frac{y+iz}{1/2-x} - i} = \frac{-(x + iy) + (1/2 + z)}{(x - iy) - (1/2 - z)}
$$

$$
= \frac{\zeta(1/2 - z) + (1/2 + z)}{\frac{1}{\zeta}(1/2 + z) - (1/2 - z)} = \zeta
\tag{6.59}
$$

From this we get the following relations between w, w', ζ and ζ':

$$
w = i\frac{\zeta + 1}{\zeta - 1} \quad \text{and} \quad w' = i\frac{\zeta' + 1}{\zeta' - 1}
\tag{6.60}
$$

and from equation 6.58

$$\frac{\zeta' + 1}{\zeta' - 1} = e^{i\theta} \frac{\zeta + 1}{\zeta - 1} \tag{6.61}$$

Solving this equation for ζ' we find:

$$\zeta' = \frac{\zeta(e^{i\theta} + 1) + (e^{i\theta} - 1)}{\zeta(e^{i\theta} - 1) + (e^{i\theta} + 1)} \tag{6.62}$$

Multiplying nominator and denominator by $(1 + e^{-i\theta})$ and using some standard trigonometrical indentities we find finally:

$$\zeta' = \frac{\zeta \cos \frac{\theta}{2} + i \sin \frac{\theta}{2}}{i\zeta \sin \frac{\theta}{2} + \cos \frac{\theta}{2}} = \sigma(\alpha, \beta, -\overline{\beta}, \overline{\alpha})\zeta \tag{6.63}$$

were we defined

$$\alpha = \cos \frac{\theta}{2} \quad \text{and} \quad \beta = i \sin \frac{\theta}{2}. \tag{6.64}$$

This theorem establishes a mapping from the set of 3-D motions to the set of 2×2 complex matrices with determinant one. The matrices $\sigma(a, b, c, d)$ and $-\sigma(a, b, c, d)$ both describe the same function $\frac{a\zeta + b}{c\zeta + d}$ and they both have determinant one. We see that a rigid motion in 3-D is related to two 2×2 matrices $\sigma(a, b, c, d)$ and $-\sigma(a, b, c, d)$ via the stereographic projection.

In the next theorem we will show that we get all 2×2 matrices with determinant one in this way.

Theorem 6.8 Let $\sigma(v, w, x, y)$ be a matrix with complex entries v, w, x and y such that $vw - xy = 1$. Then it is possible to find a real number α and complex numbers β, a and b such that

$$\begin{pmatrix} v & w \\ x & y \end{pmatrix} = \begin{pmatrix} \alpha & \beta \\ 0 & 1/\alpha \end{pmatrix} \begin{pmatrix} a & b \\ -\overline{b} & \overline{a} \end{pmatrix} \tag{6.65}$$

Furthermore α, β, a and b are uniquely defined by v, w, x and y up to a common sign.

This means especially that for each matrix $\sigma(v, w, x, y)$ we can find a translation h and a rotation g such that the rigid motion defined by g and h is mapped to the matrix $\sigma(v, w, x, y)$ under the stereographic projection.

To show the equality in equation 6.65 we rewrite the equation in the four separate component equations:

$$v = a\alpha - \overline{b}\beta, \quad w = b\alpha + \overline{a}\beta, \quad x = -\frac{\overline{b}}{\alpha} \quad y = \frac{\overline{a}}{\alpha} \tag{6.66}$$

Calculating a and b from these equations we find:

$$a = \overline{y}\alpha \quad \text{and} \quad b = -\overline{x}\alpha \tag{6.67}$$

Inserting this in the equations 6.66 gives:

$$v = \overline{y}\alpha^2 + x\alpha\beta \quad \text{and} \quad w = -\overline{x}\alpha^2 + y\alpha\beta. \tag{6.68}$$

or

$$vy = |y|^2 \alpha^2 + xy\alpha\beta \quad \text{and} \quad wx = |x|^2 \alpha^2 + xy\alpha\beta. \tag{6.69}$$

Combining the two last equations we find

$$1 = vy - wx = \alpha^2 \left(|x|^2 + |y|^2 \right) \tag{6.70}$$

from which we find that α is up to the sign uniquely defined by the original matrix $\begin{pmatrix} v & w \\ x & y \end{pmatrix}$. Using the value for α in the equations in 6.67 we find that also a and b are up to a sign uniquely defined by this matrix. Inserting the values for a, b and α in the equations 6.68 shows that also β can be obtained this way.

Before we continue with the analysis of the general case we will in the next section summarize some results from the theory of scale-invariant pattern recognition. This will then serve as a guideline when we establish some results for the general case.

6.6.3 The Mellin Transform

Two especially simple types of motion are given by camera rotations around the optical axis and by movements along the optical axis. In the first case the detected images form 2-D rotation-invariant classes and in the second case we get 2-D scale-invariant pattern classes. The problem of rotation-invariant pattern recognition was investigated earlier and we will now derive some basic results for the scale-invariant case.

We write the prototype pattern p_0 again as a function in the polar coordinate system: $p_0 = p_0(r, \varphi)$. The symmetry group is $\mathbf{R}^+$, the positive real numbers under multiplication. If the group element $g \in \mathbf{R}^+$ describes scaling with the factor s then the transformation rule is:

$$p(r, \varphi) = p_0^g(r, \varphi) = p_0(r, \varphi, s) = p_0(rs, \varphi)$$

The patterns in one such scale-invariant pattern class are functions of the polar coordinates (r, φ) and the scale variable s. We write again: $p_0^g(r, \varphi) = p_0(r, \varphi, s)$ where s is the scale factor belonging to the transformation g.

The Haar integral of the scaling group is given by:

$$\int_0^\infty f(s) \, d\mu(s) = \int_0^\infty f(s) \, \frac{d(s)}{s}$$

In the case of the scaling group $\mathbf{R}^+$ the matrix entries of the group representations are $e^{iu \ln s} = s^{iu}$ for a real number u and the weighted group averages become thus:

$$\begin{aligned}
\int_0^\infty p_0(r, \varphi, s) s^{iu-1} \, ds &= \int_0^\infty p_0(rs, \varphi) s^{iu-1} \, ds \\
&= r^{iu} \int_0^\infty p_0(s, \varphi) s^{iu-1} \, ds \\
&= r^{-iu} M_{p_0}(u, \varphi)
\end{aligned}$$

where $M_{p_0}(u, \varphi)$ is the *Mellin transform* of the prototype pattern at the point u. For further work on the Mellin transform in pattern recognition the reader may consult the literature (for example [42] and [43]).

6.6.4 Fourier Analysis of Camera Motion

In the previous sections it was shown that there is a close relation between the group of rigid motions in three-dimensional space and the group of fractional linear transformations in the complex plane. We will now sketch a theory that will allow us to analyze this type of image transformation. We will motivate the concepts heuristically by generalizing the Mellin transform described in the previous section. A full description of these theories is beyond the scope of these notes. For a detailed treatment the reader may consult the literature (see for example [36] and [14]).

We saw that under a camera motion the image point ζ moved to the point $\frac{\alpha\zeta+\beta}{\gamma\zeta+\delta}$ and the images before and after the rigid motion g are thus given by $p(\zeta)$ and

$$p^g(\zeta) = p(\tfrac{\alpha\zeta+\beta}{\gamma\zeta+\delta})$$

respectively. This means that brightness changes only due to motion. Given a fixed view $p_0(z)$ of a scene we can then define the *motion-invariant image class* p_0^G as:

$$p_0^G = \left\{ p_0^g(\zeta) = p_0(\zeta, g) = p_0(\tfrac{\alpha\zeta+\beta}{\gamma\zeta+\delta}) : \begin{pmatrix} \alpha & \gamma \\ \beta & \delta \end{pmatrix} \in \mathrm{SL}(2,\mathbf{C}) \right\}$$

Our goal is now to find an expression for the weighted group averages for this pattern class. For notational convenience we will ignore the spatial variable ζ in the following. We consider $p_0(\zeta, g)$ as a function of the group elements g only and write thus $p_0(g)$ instead of $p_0(\zeta, g)$.

We consider the group invariant class of patterns p_0^G. The Haar integral on $\mathrm{SL}(2,\mathbf{C})$ is given by: $\int \varphi(g)\, dg$ with

$$dg = \left(\frac{i}{2}\right)^3 |\beta|^{-2}\, d\alpha\, d\overline{\alpha}\, d\beta\, d\overline{\beta}\, d\gamma\, d\overline{\gamma}$$

where g is the matrix $\begin{pmatrix} \alpha & \gamma \\ \beta & \delta \end{pmatrix}$.

We saw in section 4.3 that the operator $T_\chi : D_\chi \to D_\chi$ defined as

$$T_\chi\varphi(z) = (\beta z + \delta)^{n_1-1}(\overline{\beta z} + \overline{\delta})^{n_2-1}\varphi\left(\frac{\alpha z+\beta}{\gamma z+\delta}\right) \tag{6.71}$$

defined a representation on the space D_χ. This space was defined as:

$$D_\chi = \{\varphi : \text{there is a } \psi \in H_\chi \text{ such that: } \varphi(z) = \psi(z_1/z_2, 1)\} \tag{6.72}$$

where H_χ was the space of homogeneous polynomials of degree $\chi = (n_1 - 1, n_2 - 1)$.

We define the *Fourier transform* of the function $p(g)$ on the group $\mathrm{SL}(2,\mathbf{C})$ as:

$$F(\chi) = \int p(g)T_\chi(g)\, dg \tag{6.73}$$

Note that this is nothing else but our old weighted group averages $(\int_G p(g)t_{kl}^n(g)\, dg$ with matrix entries $t_{kl}^n)$ in compact operator notation. Under certain conditions on the function p (p must be summable) it is known that $F(\chi)$ exists for all χ with $-2 \leq \mathrm{Re}(n_1 + n_2) \leq 2$.

T is a representation and $T(g)$ is thus a matrix describing an operator on D_χ. $p(g)$ is a complex value and $F(\chi)$ is thus an *operator* on D_χ. It can be shown that $F(\chi)$ acts as an integral operator with a kernel $K(z_1, z_2; \chi)$ for all functions $\varphi \in D_\chi$:

$$F(\chi)\varphi(z_1) = \int p(g)T_\chi(g)\varphi(z_1)\, dg = \frac{i}{2} \int K(z_1, z_2; \chi)\varphi(z_2)\, dz_2 d\overline{z_2} \qquad (6.74)$$

If we describe $g \in \mathrm{SL}(2,\mathbf{C})$ in the three variables α, β and δ then p is a function $p(g) = p(\alpha, \beta, \delta)$. For the kernel we then find the following expression:

$$K(z_1, z_2, \chi) = \left(\frac{i}{2}\right)^2 \int p(\lambda^{-1} + \beta z_2, \beta, \lambda - \beta z_1)\lambda^{n_1-1}\overline{\lambda}^{n_2-1}\, d\lambda d\overline{\lambda} d\beta d\overline{\beta} \qquad (6.75)$$

Note that this formula is very similar to the Mellin transform.

This new Fourier transform has many of the well known properties of the usual Fourier transform. Especially we have the following inverse Fourier transform:

$$
\begin{aligned}
p(g) &= p(\alpha, \beta, \delta) = \frac{1}{34\pi^4} \sum_{n=-\infty}^{\infty} \frac{i}{2} \int_{-\infty}^{\infty} (n^2 + \rho^2)\, d\rho \\
&\times \int K\left(z, \frac{\alpha z + \beta}{\gamma z + \delta}; \left(\frac{n + i\rho}{2}, \frac{-n + i\rho}{2}\right)\right) \\
&\times (\beta z + \delta)^{-\frac{1}{2}(n+i\rho)-1}(\overline{\beta}\overline{z} + \overline{\delta})^{-\frac{1}{2}(n-i\rho)-1}\, dz d\overline{z}
\end{aligned}
\qquad (6.76)
$$

There is also a Plancherel formula, i.e. we can introduce a scalar product in the space of the kernel functions $K(z_1, z_2, \chi)$ so that the Fourier Transform preserves the scalar product:

$$\langle p_1, p_2 \rangle = \langle K_1(z_1, z_2; \chi), K_2(z_1, z_2; \chi) \rangle \qquad (6.77)$$

6.7 Neural Networks

Group theoretical models of brain functions are among the earliest studies in the theory of neural networks. In their 1947 paper (see [39]) on how we know universals, Pitts and McCulloch describe the recognition of patterns independently of some type of deformation as follows:

This example may be straightforwardly generalized to provide a uniform principle of design for reflex-mechanisms which secure invariance under an arbitrary group G. In some way, out of a whole series of transforms $T\phi$ of an apparition, one of them ϕ_0 is elected to be standard - e.g., one of a standard overall size - and when presented with ϕ, the mechanism computes one or more suitable parameters $a(\phi), b(\phi), ...$, which define its position within the series of $T\phi$'s in a univocal way ...

For any general group of the type we are considering, quantities $a(\phi)$ of this type may always be found, as is shown in the theory of irreducible representations of the group G.

In later applications of group theory to neural networks (for example in [35]) it is always assumed that the group is finite. Although the number of neurons is finite it is not true that the transformation group is finite. In our study of group theoretical methods we nearly ignored finite groups and we will therefore illustrate the application of group theoretical methods in neural network theory in an example that is similar to the ones described in the Pitts and McCulloch paper mentioned above.

In our description we follow the investigation of basic units as described in [24]. In this book Kohonen introduces basic units as models of cells, or assemblies of cells, that act as linear filters. As an example consider a cell in the visual system that processes input from a limited region of the retina. The input of the cell describes the gray value distribution in this region. This region is called the the receptive field of the cell. The cell responds to this input by sending a signal to cells in the next layer of the network. Although the outgoing signal is in general a nonlinear function of the input it is interesting to approximate the response by a linear combination of the input signals. This simplifies the analysis and such processes may also be useful in the design and analysis of synthetic networks.

One of the most interesting properties of such a basic unit in a network is the ability to change its behavior dynamically. In the case of a linear unit this means that the weights used to compute the output from the incoming signals are dynamically changed. Using a learning procedure the cell tries to achieve the desired relation between the input and the output signals.

In our description of such a basic unit we assume that the possible input signals of this cell form a Hilbert space H. From an incoming signal s the unit computes the output o as the scalar product $\langle s, f \rangle = o$ where the function f can be thought as a description of the state of the unit.

A main feature of such a basic unit is the fact that it adapts itself to its environment. The transformation rule described by the element f is therefore a function of time: $f = f(t)$. The output at time t becomes thus: $o(t) = \langle s, f(t) \rangle$. We introduce now the learning rule which describes how the system changes its behavior over time. We will in the following assume that the evolution of the system is described by the differential equation:

$$f'(t) = o(t) \cdot s - \eta(t) f(t) \tag{6.78}$$

Here f' denotes the differentiation with respect to the time variable t and η is some time-varying function. Ignoring the negative term for a moment we see that a large value of $o(t)$, i.e. a high correlation between the signal s and the current filter function $f(t)$, pushes the current filter function further in the direction of the signal s. A low value of $o(t)$, indicating a small correlation between filter and pattern has little effect in the updating. The second term $\eta(t) f(t)$ can be thought of as a type of forgetting. Learning rules of this type are usually called Hebb-type learning rules.

We will now investigate how such a basic unit reacts to regularities in the space of input signals. We will use the same type of model as in the filter design example, with the exception that the filter function f now varies over time. We assume again that the space of input signals is a symmetrical signal space (H, G, T) with a Hilbert space H, a symmetry group G and a group representation T. Two signals s_1 and s_2 are equivalent if there is a group element $g \in G$ such that $s_2 = T(g)s_1$.

We will now compute the stable states f of such a basic unit, i.e. those states that are no longer changed. Substitution of $o(t)$ into equation 6.78 gives:

$$f'(t) = \langle s, f(t) \rangle \cdot s - \eta(t) f(t) \tag{6.79}$$

This updating rule is probably not invoked for each new input signal s. Instead we have to think of f as a slowly varying function which reacts only when it has seen a large number of input signals s. It thus seems reasonable to assume that f depends only the statistical properties of the input signals. In our analysis we assume that we have an expectation operator E_s that computes the average value of some function of the input elements s. We apply this operator to both sides of equation 6.79 to find:

$$E_s(f'(t)) = E(\langle s, f(t) \rangle \cdot s - \eta(t)f(t)) = E_s(\langle g, f(t) \rangle \cdot s) - \eta(t)f(t) \qquad (6.80)$$

If f is a stable solution, i.e. a solution which does not change over time then we have $f_0'(t) = 0$. This solution is independent of t and we write f_0 instead of $f_0(t)$. For f_0 we find:

$$0 = E_s(\langle s, f_0 \rangle \cdot s) - \eta(t)f_0 \qquad (6.81)$$

or

$$E_s(\langle s, f_0 \rangle \cdot s) = \eta(t)f_0 \qquad (6.82)$$

If we define the operator A as $Af = E_s(\langle s, f(t) \rangle \cdot s)$ then equation 6.82 means that f is a solution of the eigenvalue problem:

$$Af = \lambda f. \qquad (6.83)$$

Now assume that the input signals are all of the form $s = p^g$ for some fixed element p and transformations $g \in G$. Assume further that all such signals are equally probable. The operator E_s is then given by the Haar integral

$$E_s(\langle s, f \rangle \cdot s) = \int_G \langle p^g, f \rangle \cdot p^g \, dg \qquad (6.84)$$

The operator A in equation 6.83 depends now also on the fixed prototype pattern p and we rewrite the last equation 6.84 as

$$A_p f = E_s(\langle s, f \rangle \cdot s) = \int_G \langle p^g, f \rangle \cdot p^g \, dg \qquad (6.85)$$

Assume that the scalar product is invariant under the group action, i.e. $\langle f_1, f_2 \rangle = \langle f_1^g, f_2^g \rangle$ for all $g \in G$. Then we find that the operator A_p intertwines:

$$A_p(f^h) = E_s(\langle s, f^h(t) \rangle \cdot s) = \int_G \langle p^g, f^h \rangle \cdot p^g \, dg$$

$$= \int_G \langle p^{gh^{-1}}, f \rangle \cdot p^g \, dg = \int_G \langle p^g, f \rangle \cdot p^{gh} \, dg = (A_p(f))^h$$

The stable states of the network are thus eigenfunctions of the intertwining operator A_p.

These results show that the stable states of a basic unit are connected to the eigenfunctions of an intertwining operator. However, we did not investigate the question whether the unit actually converges to this solution. We will now describe an algorithm and some simulations that show how we can ensure that the unit actually reaches one of these stable solutions.

In the rest of this section we restrict our attention to the case where the scalar product in the Hilbert space is given by an integral of the form:

$$\langle f_1, f_2 \rangle = \int_{\mathbf{R}^n} f_1(x) f_2(x) m(x) \, dx \qquad (6.86)$$

where $m(x)$ is a fixed weight function and the integral $\int ...dx$ is the Lebesgue integral. The operator A_p becomes in this case:

$$(A_p f)(y) = \int_G \langle p^g, f \rangle \, p^g \, dg = \int_G \int_{\mathbf{R}^n} p^g(x) f(x) m(x) \, dx p^g(y) \, dg. \qquad (6.87)$$

Changing the integration order gives:

$$(A_p f)(y) = \int_{\mathbf{R}^n} \int_G p^g(x) p^g(y) \, dg f(x) m(x) \, dx = \int_{\mathbf{R}^n} k_p(x,y) f(x) \, dx \qquad (6.88)$$

with $k_p(x,y) = \int_G p^g(x) p^g(y) m(x) \, dg$. The operator A_p is thus an integral operator with the kernel k_p. If the operator k_p is known then one can show that the series $f_0, f_1, ...$ with:

$$f_{l+1} = \frac{1}{\|A_p f_l\|} A_p f_l \qquad (6.89)$$

converges to an eigenfunction of A_p independent of the start function f_0.

In our application k_p is not known, however; all we have is a series of signals p_l of the form $p_l = p^{g_l}$ for certain group elements g_l. Instead of using the true kernel k_p we iterate thus with the approximate kernel

$$k_p^l(x,y) = \frac{m(x)}{l} \sum_{\nu=1}^{l} p^{g_l}(x) p^{g_l}(y).$$

The associated operator A_p^l is then given by:

$$(A_p^l f)(y) = \int_{\mathbf{R}^n} k_p^l(x,y) f(x) \, dx$$

The corresponding iteration rule is:

$$f_{l+1} = \frac{1}{\|A_p^l f_l\|} A_p f_l \qquad (6.90)$$

In the following example we consider signals f that are defined on the unit circle. In image processing these signals can represent circular patterns whose gray value distribution is independent of the radius. Typical examples of such patterns are step edges or lines. A 2-D rotation transforms a pattern by shifting it some distance to the left or the right. The following two diagrams 6.11 show two such signals that could represent two step edges with different orientations. Each pattern consists of 256 points.

In the next figure 6.12 we see the function f_{2000}, i.e. the state of the basic unit after 2000 iterations. For comparison we show also the cosine and the sine function in the same plot (all functions f, including the sine and cosine, are normed so that $\|f\| = 1$). The functions $\cos x$ and $\sin x$ belong to the same eigenvalue and therefore we find that all functions of the form $a \sin x + b \cos x$ with $a^2 + b^2 = 1$ are also eigenfunctions with the same eigenvalue. Especially all functions $\cos(x + \gamma)$ with a fixed constant γ are eigenfunctions. In the previous example we see that the unit did stabilize in one of these shifted functions $\cos(x + \gamma)$ with $\gamma \neq 0$.

In the next two figures we show first (figure 6.13) two typical line-like input signals and then in figure 6.14 the resulting filter function. The signals consisted again of 256

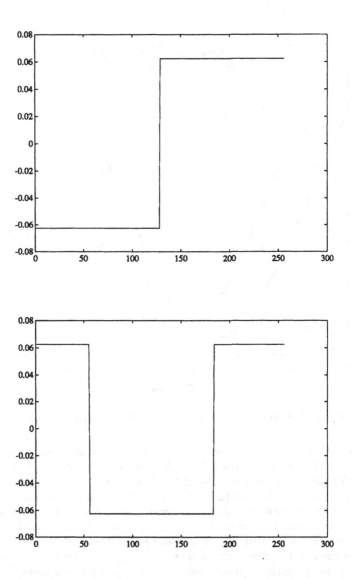

Figure 6.11: Two different edge-like input signals

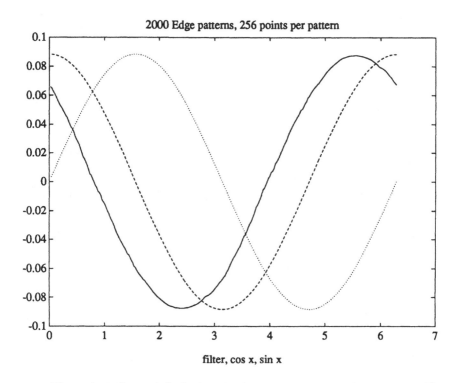

Figure 6.12: State of the basic unit after 2000 iterations using operators k^l

points each. These experiments show that the basic unit stabilizes in states of the form $\gamma \cos(2n + 1)x + (1 - \gamma) \sin(2n + 1)x$ if the prototype was odd (as in the edge-detection example) and it stabilizes in states of the form $\gamma \cos 2nx + (1 - \gamma) \sin 2nx$ if the input pattern was an even function (as in the line-detection example), where γ is a constant. A similar result for the design of rotation-invariant filter functions was derived in [28].

This implementation of Hebb learning is of course computationally very expensive since the covariance matrix (i.e. the operator k^l) must be available all the time. However, we can simplify the computations considerably by going back to the original type of learning rule as described in equation 6.79. An easy calculation shows that the previous update rule involving the covariance matrix is equivalent to the following equation:

$$f(n + 1) = \frac{1}{n} o(n)p(n) + f(n) \tag{6.91}$$

where $f(n)$ is the filter function at the discrete time-step n, $o(n)$ is the output and $p(n)$ is the pattern at time-step n. The next two figures (figure 6.15) show that this algorithm leads to the same stable states of the unit.

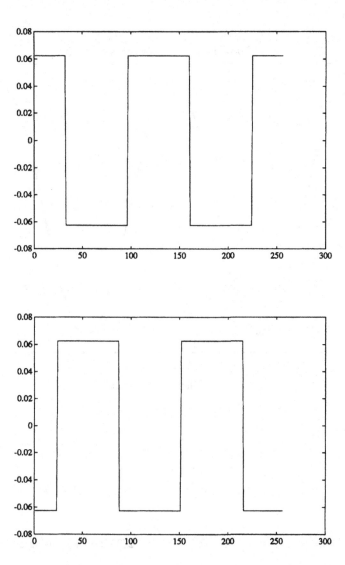

Figure 6.13: Two line-like angular input signals

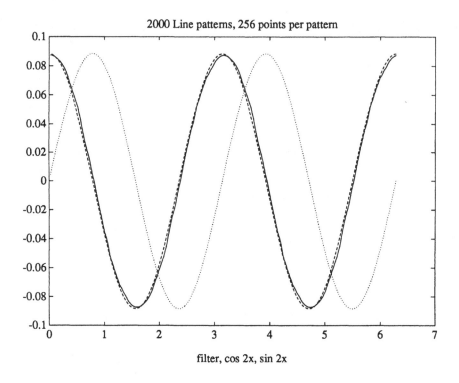

Figure 6.14: State of the basic unit after 2000 iterations using operators k^l

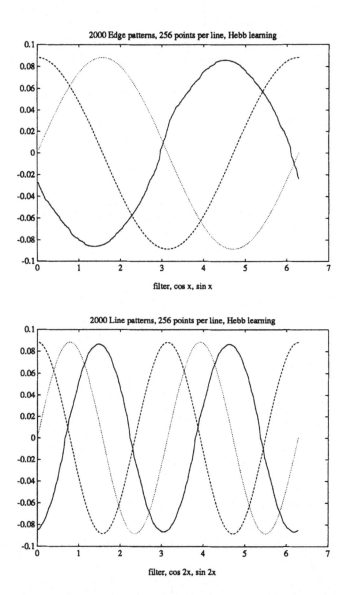

Figure 6.15: State of the basic unit after 2000 iterations using edges and lines

Bibliography

[1] H. Arsenault and C. Delisle. Contrast-invariant pattern recognition using circular components. *Applied Optics*, 24(14):2072–2075, 1985.

[2] H. Arsenault and Y. Sheng. Modified composite filter for pattern recognition in the presence of noise with non-zero mean. *Optics Communications*, 63(1):15–20, 1987.

[3] J. Babaud, A. P. Witkin, M. Baudin, and R. O. Duda. Uniqueness of the gaussian kernel for scale-space filtering. *IEEE Transactions on Pattern Analysis and Machine Intelligence*, PAMI-8(1):26–33, 1986.

[4] L. Breiman. *Probability.* Addison-Wesley, Reading, 1968.

[5] C. Chevalley. *Theory of Lie-Groups.* Princeton University Press, 1946.

[6] P. E. Danielsson. Rotation invariant linear operators with directional response. In *Proc. 5. ICPR*, pages 1171–1176, 1980.

[7] P. E. Danielsson. Natural basis functions for image analysis. In D. Meyer-Ebrecht, editor, *Proceedings of the 6. Aachener Symposium für Signaltheorie*, Informatik-Fachberichte, Vol. 153, pages 239–254, Springer-Verlag, Berlin, Heidelberg, New York, 1987.

[8] P. E. Danielsson and Henrik Sauleda. Rotation invariant 2d filters matched to 1d features. In *Proceedings of the IEEE Computer Society Conference on Computer Vision and Pattern Recognition, San Francisco*, 1985.

[9] S. R. Deans. *The Radon Transform and Some if its Applications.* Wiley Interscience, 1983.

[10] N. Dunford and J. T. Schwartz. *Linear Operators (Vols. I, II).* Interscience Publishers, New York, 1958, 1963.

[11] A. Erdelyi, W. Magnus, F. Oberhettinger, and F. G. Tricomi. *Higher Transcendental Functions.* McGraw-Hill, New York, Toronto, London, 1953.

[12] Rumelhart et al. Learning representation by back-propagation errors. *Nature*, 323(6088):5533, 1986.

[13] Rumelhart et al. *Parallel Distributed Processing (Vols. 1,2).* MIT Press, Cambridge, 1986.

[14] I. M. Gelfand, M. I. Graev, and N. Y. Vilenkin. *Generalized Functions, Vol. 5 (Integral Geometry and Representation Theory)*. Academic Press, 1966.

[15] I. M. Gelfand, R. A. Minlos, and Z. Y. Shapiro. *Representations of the rotation and Lorentz groups and their applications*. Pergamon Press, 1963.

[16] J. Goodman. *Introduction to Fourier Optics*. Physical and Quantum Electronics Series. McGraw-Hill, 1968.

[17] P.R. Halmos, editor. *Measure Theory*. D. Van Nostrand, Reading, Mass., 1950.

[18] H.D.Ebbinghaus, H. Hermes, F. Hirzebruch, M. Koecher, K. Mainzer, J. Neukirch, A. Prestel, and R. Remmert. *Zahlen*. Springer-Verlag, Berlin, Heidelberg, New York, 1988.

[19] F. Hirzebruch, W. Scharlau. *Einführung in die Funktionalanalysis*. Bibliographisches ches Institut, Mannheim, Wien, Zürich, 1971.

[20] R. A. Hummel. Feature detection using basis functions. *Computer Graphics and Image Processing*, 9:40–55, 1979.

[21] S. H. Izen. A series inversion for the x-ray transform in n dimensions. *Inverse Problems*, 4:725–748, 1988.

[22] K. I. Kanatani. Camera rotation invariance of image characteristics. *Computer Vision, Graphics and Image Processing*, 39:328–354, 1987.

[23] K. I. Kanatani. Transformation of optical flow by camera rotation. *IEEE Transactions on Pattern Analysis and Machine Intelligence*, 10(2):131–143, 1988.

[24] T. Kohonen. *Self organization and associative memory*. Springer Series in Information Sciences, Vol. 8. Springer-Verlag, Berlin, Heidelberg, New York, 1984.

[25] S. Lang. *Algebra*. Addison-Wesley, 1965.

[26] G. Laub and R. Bachus. Visualization of vessels with mri. In *Proc. Computer Aided Radiology*. Springer-Verlag, Berlin, Heidelberg, New York, 1987.

[27] R. Lenz. *Reconstruction, Processing and Display of 3D-Images*. PhD thesis, Linköping University, Sweden, 1986.

[28] R. Lenz. Optimal filters for the detection of linear patterns in 2-d and higher dimensional images. *Pattern Recognition*, 20(2):163–172, 1987.

[29] R. Lenz. Rotation-invariant operators and scale-space filtering. *Pattern Recognition Letters*, 6:151–154, 1987.

[30] R. Lenz. A group theoretical approach to filter design. In *Proc. International Conference on Accoustics, Speech and Signal Processing*, 1989.

[31] R. Lenz. A group theoretical model of feature extraction. *Journal of the Optical Society of America A*, 6(6):827–834, June 1989.

[32] M. P. Levesque and H. H. Arsenault. Rotation-invariant pattern recognition using the phase of the circular harmonic filter correlations. *Optics Communications*, 58(3):161–166, 1986.

[33] R. Lippman. An introduction to computing with neural nets. *IEEE ASSP Magazine*, 4:4–22, 1987.

[34] D. Marr. *Vision*. W. H. Freeman, 1982.

[35] M. Minsky and S. Papert. *Perceptrons*. MIT Press, Cambridge, 1969.

[36] M. A. Naimark. *Linear Representations of the Lorentz Group*. Pergamon Press, Oxford, London, 1964.

[37] M. A. Naimark and A. I. Stern. *Theory of Group Representations*. Springer-Verlag, New York, Berlin, Heidelberg, 1982.

[38] A. Papoulis. *System and Transforms with Applications in Optics*. McGraw-Hill, New York, 1968.

[39] W. Pitts and W. S. McCulloch. How we know universals: the perception of auditory and visual forms. *Bulletin of Mathematical Biophysics*, 9:127–147, 1947.

[40] L. Pontrjagin. *Topological Groups*. Princeton Mathematical Series No. 2, 1946.

[41] F. Riesz and B. Sz. Nagy. *Vorlesungen über Funktionalanalysis*. Deutscher Verlag der Wissenschaften, Berlin, 1956.

[42] Y. Sheng and H. H. Arsenault. Experiments on pattern recognition using invariant fourier- mellin descriptors. *Journal of the Optical Society of America A*, 3(6):771–776, 1986.

[43] Y. Sheng and J. Duvernoy. Circular-fourier-radial-mellin transform descriptors for pattern recognition. *Journal of the Optical Society of America A*, 3(6):885–888, 1986.

[44] D. Slepian. Prolate spheroidal wave functions, fourier analysis and uncertainty - iv: Extensions to many dimensions; generalized prolate spheroidal functions. *Bell System Technical Journal*, 43:3009–3058, 1964.

[45] M. Sugiura. *Unitary Representations and Harmonic Analysis*. Kodansha Ltd., Tokyo, 1975.

[46] A. Terras. *Harmonic Analysis on Symmetric Spaces and Applications I*. Springer-Verlag, Berlin, Heidelberg, New York, 1985.

[47] V. S. Varadarajan. *Lie Groups, Lie Algebras and their Representations*. Prentice-Hall, Englewood Cliffs, N.J., 1974.

[48] N. Ya. Vilenkin. *Special Functions and the Theory of Group Representations*. AMS Translations. American Mathematical Society, Providence, R.I., 1968.

[49] V. S. Vladimirov. *Equations of Mathematical Physics*. Mir Publishers, Moscow, 1984.

[50] G. N. Watson. *A Treatise on the Theory of Bessel Functions*. Cambridge University Press, 1966.

[51] R. Wu and H. Stark. Rotation-invariant pattern recognition using a vector reference. *Applied Optics*, 23(6):838–840, 1984.

[52] R. Wu and H. Stark. Rotation-invariant pattern recognition using optimum feature extraction. *Applied Optics*, 24(2):179–184, 1985.

[53] K. Yosida. *Functional Analysis*. Springer-Verlag, Berlin, Heidelberg, New York, 1978.

[54] A. L. Yuille and T. Poggio. Scaling theorems for zero crossings. *IEEE Transactions on Pattern Analysis and Machine Intelligence*, PAMI-8(1):15–25, 1986.

[55] D. P. Zelobenko. *Compact Lie Groups and their Representations*. American Mathematical Society, Providence, R. I., 1973.

[56] S. W. Zucker and R. A. Hummel. A three-dimensional edge operator. *IEEE Transactions on Pattern Analysis and Machine Intelligence*, PAMI-3(3):324–331, 1981.

Index

List of Figures

Vol. 379: A. Kreczmar, G. Mirkowska (Eds.), Mathematical Foundations of Computer Science 1989. Proceedings, 1989. VIII, 605 pages. 1989.

Vol. 380: J. Csirik, J. Demetrovics, F. Gécseg (Eds.), Fundamentals of Computation Theory. Proceedings, 1989. XI, 493 pages. 1989.

Vol. 381: J. Dassow, J. Kelemen (Eds.), Machines, Languages, and Complexity. Proceedings, 1988. VI, 244 pages. 1989.

Vol. 382: F. Dehne, J.-R. Sack, N. Santoro (Eds.), Algorithms and Data Structures. WADS '89. Proceedings, 1989. IX, 592 pages. 1989.

Vol. 383: K. Furukawa, H. Tanaka, T. Fujisaki (Eds.), Logic Programming '88. Proceedings, 1988. VII, 251 pages. 1989 (Subseries LNAI).

Vol. 384: G. A. van Zee, J. G. G. van de Vorst (Eds.), Parallel Computing 1988. Proceedings, 1988. V, 135 pages. 1989.

Vol. 385: E. Börger, H. Kleine Büning, M. M. Richter (Eds.), CSL '88. Proceedings, 1988. VI, 399 pages. 1989.

Vol. 386: J.E. Pin (Ed.), Formal Properties of Finite Automata and Applications. Proceedings, 1988. VIII, 260 pages. 1989.

Vol. 387: C. Ghezzi, J. A. McDermid (Eds.), ESEC '89. 2nd European Software Engineering Conference. Proceedings, 1989. VI, 496 pages. 1989.

Vol. 388: G. Cohen, J. Wolfmann (Eds.), Coding Theory and Applications. Proceedings, 1988. IX, 329 pages. 1989.

Vol. 389: D. H. Pitt, D. E. Rydeheard, P. Dybjer, A. M. Pitts, A. Poigné (Eds.), Category Theory and Computer Science. Proceedings, 1989. VI, 365 pages. 1989.

Vol. 390: J.P. Martins, E.M. Morgado (Eds.), EPIA 89. Proceedings, 1989. XII, 400 pages. 1989 (Subseries LNAI).

Vol. 391: J.-D. Boissonnat, J.-P. Laumond (Eds.), Geometry and Robotics. Proceedings, 1988. VI, 413 pages. 1989.

Vol. 392: J.-C. Bermond, M. Raynal (Eds.), Distributed Algorithms. Proceedings, 1989. VI, 315 pages. 1989.

Vol. 393: H. Ehrig, H. Herrlich, H.-J. Kreowski, G. Preuß (Eds.), Categorical Methods in Computer Science. VI, 350 pages. 1989.

Vol. 394: M. Wirsing, J.A. Bergstra (Eds.), Algebraic Methods: Theory, Tools and Applications. VI, 558 pages. 1989.

Vol. 395: M. Schmidt-Schauß, Computational Aspects of an Order-Sorted Logic with Term Declarations. VIII, 171 pages. 1989. (Subseries LNAI).

Vol. 396: T. A. Berson, T. Beth (Eds.), Local Area Network Security. Proceedings, 1989. IX, 152 pages. 1989.

Vol. 397: K. P. Jantke (Ed.), Analogical and Inductive Inference. Proceedings, 1989. IX, 338 pages. 1989. (Subseries LNAI).

Vol. 398: B. Banieqbal, H. Barringer, A. Pnueli (Eds.), Temporal Logic in Specification. Proceedings, 1987. VI, 448 pages. 1989.

Vol. 399: V. Cantoni, R. Creutzburg, S. Levialdi, G. Wolf (Eds.), Recent Issues in Pattern Analysis and Recognition. VII, 400 pages. 1989.

Vol. 400: R. Klein, Concrete and Abstract Voronoi Diagrams. IV, 167 pages. 1989.

Vol. 401: H. Djidjev (Ed.), Optimal Algorithms. Proceedings, 1989. VI, 308 pages. 1989.

Vol. 402: T. P. Bagchi, V. K. Chaudhri, Interactive Relational Database Design. XI, 186 pages. 1989.

Vol. 403: S. Goldwasser (Ed.), Advances in Cryptology – CRYPTO '88. Proceedings, 1988. XI, 591 pages. 1990.

Vol. 404: J. Beer, Concepts, Design, and Performance Analysis of a Parallel Prolog Machine. VI, 128 pages. 1989.

Vol. 405: C. E. Veni Madhavan (Ed.), Foundations of Software Technology and Theoretical Computer Science. Proceedings, 1989. VIII, 339 pages. 1989.

Vol. 407: J. Sifakis (Ed.), Automatic Verification Methods for Finite State Systems. Proceedings, 1989. VII, 382 pages. 1990.

Vol. 408: M. Leeser, G. Brown (Eds.) Hardware Specification, Verification and Synthesis: Mathematical Aspects. Proceedings, 1989. VI, 402 pages. 1990.

Vol. 409: A. Buchmann, O. Günther, T. R. Smith, Y.-F. Wang (Eds.), Design and Implementation of Large Spatial Databases. Proceedings, 1989. IX, 364 pages. 1990.

Vol. 410: F. Pichler, R. Moreno-Diaz (Eds.), Computer Aided Systems Theory – EUROCAST '89. Proceedings, 1989. VII, 427 pages. 1990.

Vol. 411: M. Nagl (Ed.), Graph-Theoretic Concepts in Computer Science. Proceedings, 1989. VII, 374 pages. 1990.

Vol. 412: L. B. Almeida, C. J. Wellekens (Eds.), Neural Networks. Proceedings, 1990. IX, 276 pages. 1990.

Vol. 413: R. Lenz, Group Theoretical Methods in Image Processing. VIII, 139 pages. 1990.

Printed in the United States
By Bookmasters